ENERGY SCIENCE, ENGINEERING AND TECHNOLOGY

U.S. AND WORLD COAL PRODUCTION

DEVELOPMENTS AND PROJECTIONS

ENERGY SCIENCE, ENGINEERING AND TECHNOLOGY

Additional books in this series can be found on Nova's website under the Series tab.

Additional E-books in this series can be found on Nova's website under the E-book tab.

ENERGY POLICIES, POLITICS AND PRICES

Additional books in this series can be found on Nova's website under the Series tab.

Additional E-books in this series can be found on Nova's website under the E-book tab.

ENERGY SCIENCE, ENGINEERING AND TECHNOLOGY

U.S. AND WORLD COAL PRODUCTION

DEVELOPMENTS AND PROJECTIONS

RAUL M. SCHMIDT
EDITOR

NOVINKA

New York

For permission to use material from this book please contact us:
Telephone 631-231-7269; Fax 631-231-8175
Web Site: http://www.novapublishers.com

Library of Congress Cataloging-in-Publication Data

ISBN: 978-1-62618-976-8

Published by Nova Science Publishers, Inc. †New York

CONTENTS

PREFACE

Even though U.S. coal production remained strong over the past decade, reaching record levels of production, coal is losing its share of overall U.S. energy production primarily to natural gas. One of the big questions for the industry is how to penetrate the overseas market, particularly in steam coal, to compensate for declining domestic demand. As U.S. energy policy and environmental regulations are constantly debated, there is ongoing congressional interest in the role of coal in meeting U.S. and global energy needs. The question may not be whether the domestic production of coal is here to stay but, rather, how much U.S. coal will be mined, what type, and under what regulatory framework. This book serves as a primer on U.S. and world coal resources and production and highlights some of the congressional interest related to coal production on U.S. federal lands. The book also describes the past ten years of coal activity and includes a discussion of future coal production projections and federal coal incentives, many of which do not expire.

Chapter 1 – Even though U.S. coal production remained strong over the past decade, reaching record levels of production, coal is losing its share of overall U.S. energy production primarily to natural gas. One of the big questions for the industry is how to penetrate the overseas market, particularly in steam coal, to compensate for declining domestic demand. As U.S. energy policy and environmental regulations are constantly debated, there is ongoing congressional interest in the role of coal in meeting U.S. and global energy needs. The question may not be whether the domestic production of coal is here to stay but, rather, how much U.S. coal will be mined, what type, and under what regulatory framework. Energy Information Administration (EIA) statistics show that more than half (55%) of U.S. coal reserves are located in the West, dominated by Montana and Wyoming, which account for 43%. When including the top five producing states (three of which are in the East), 70% of U.S. coal reserves are accounted for. The United States government owns about one third, or 87 billion short tons (BST), of U.S. domestic reserves. Coal production in the United States reached an all-time high of 1,174.8 million short tons in 2008, before declining to slightly under 1,100 million short tons from 2009 to 2011. Coal production on federal lands accounts for about 43% of U.S. production, according to the Bureau of Land Management (BLM). World coal production has increased by nearly 60% since 2002, most of the increase coming from China—up 130%. Overall, U.S. coal production has been very strong over the past decade and if the industry is successful in penetrating the global market, primarily for steam coal, U.S. production may continue to grow faster than consumption. If recent trends continue, the U.S. coal industry will likely become more concentrated and produce more on federal lands. Businesses in the coal industry are subject to federal income taxes, but coal producers benefit from a number of federal tax provisions, commonly referred to as tax expenditures. There are several congressional concerns related to coal production on federal lands. One concern is the potential for under-market-value coal auctions (sales), e.g., lease offers being accepted by the BLM with few competitive bids. In these cases, the federal government may not receive fair market value for the lease sale.

Chapter 2 – Mountaintop removal mining involves removing the top of a mountain in order to recover the coal seams contained there. This practice occurs in six Appalachian states (Kentucky, West Virginia, Virginia, Tennessee, Pennsylvania, and Ohio). It creates an immense quantity of excess spoil (dirt and rock that previously composed the mountaintop), which is typically placed in valley fills on the sides of the former mountains, burying streams that flow through the valleys. Mountaintop mining is regulated under several laws, including the Clean Water Act (CWA) and the Surface Mining Control and Reclamation Act (SMCRA). Critics say that, as a result of valley fills from mountaintop mining, stream water quality and the aquatic and wildlife habitat that streams support are destroyed by tons of rocks and dirt. The mining industry argues that mountaintop mining is essential to conducting surface coal mining in the Appalachian region and that it would not be economically feasible there if operators were barred from using valleys for the disposal of mining overburden. Critics have used litigation to challenge the practice. In a number of cases discussed in this report, environmental groups have been successful at the federal district court level in challenging issuance of permits for mountaintop mining projects, but each has been later overturned on appeal. Nonetheless, the criticisms also have prompted some regulatory changes, also discussed here. In June 2009, officials of the Environmental Protection Agency (EPA), the U.S. Army Corps of Engineers (Corps), and the Department of the Interior signed a Memorandum of Understanding outlining a series of administrative actions under these laws to reduce the harmful environmental impacts of mountaintop mining and surface coal mining in Appalachia. The plan includes a series of near-term and longer-term actions that emphasize specific steps, improved coordination, and greater transparency of decisions. The actions are being implemented through regulatory proposals, guidance documents, and review of applications for permits to authorize surface coal mining operations in Appalachia. Viewed broadly, the Administration's combined actions on mountaintop mining displease both industry and environmental advocates. The additional scrutiny of permits and more stringent requirements have angered the coal industry and many of its supporters. Controversy also was generated by EPA's January 2011 veto of a CWA permit that had been issued by the Corps for a surface coal mining project in West Virginia. At the same time, while environmental groups support EPA's steps to restrict the practice, many favor tougher requirements or even total rejection of mountaintop mining in Appalachia. Federal courts have recently rejected several of the Administration's actions, including overturning EPA's veto of the West Virginia mine permit, enhanced permit review procedures, and EPA guidance on factors used in evaluating water quality impacts of Appalachian surface mining permits. This report provides background on regulatory requirements, controversies and legal challenges to mountaintop mining, and recent Administration actions. Congressional interest in these issues also is discussed, including legislation in the 111th Congress seeking to restrict the practice of mountaintop mining and other legislation intended to block the Obama Administration's regulatory actions. Attention to EPA's veto of the West Virginia mining permit and other federal agency actions has increased in the 112th Congress. Several bills have been introduced to clarify or restrict EPA's authority to veto CWA permits issued by the Corps (H.R. 457/S. 272; H.R. 517; H.R. 960/S. 468; and H.R. 2018, which the House passed in July 2011).

Chapter 3 – For most of the twentieth century, the primary use of coal in the United States was for electric power generation, and for most of the history of power generation in the United States, coal has been the dominant fuel used to produce electricity. Even as recently as 2011, coal was the fuel used for almost 42% of power generation in the United States accounting for 93% of coal use. Industrial uses represented the remaining 7%. However, in April

2012, coal's share of the power generation market dropped to about 32% (according to Energy Information Administration statistics), equal to that of natural gas. Coal was the fuel of choice because of its availability and the relatively low cost of producing electricity in large, coal-burning power plants which took advantage of coal's low-priced, high energy content to employ economies of scale in steam-electric production. However, coal use for power generation seems to be on the decline, and the magnitude of coal's role for power generation is in question. Two major reasons are generally seen as being responsible: the expectation of a dramatic rise in natural gas supplies, and the impact of environmental regulations on an aging base of coal-fired power plants. A recent drop in natural gas prices has been enabled by increasing supplies of natural gas largely due to horizontal drilling and hydraulic fracturing (i.e., fracking) of shale gas formations. If the production can be sustained in an environmentally acceptable manner, then a long-term, relatively inexpensive supply of natural gas could result. Decreased natural gas prices are lowering wholesale electricity prices, stimulating a major switch from coal to gas-burning facilities. The electric utility industry values diversity in fuel choice options since reliance on one fuel or technology can leave electricity producers vulnerable to price and supply volatility. However, an "inverse relationship" may be developing for coal vs. natural gas as a power generation choice based on market economics alone, and policies which allow one fuel source to dominate may come at the detriment of the other. Coal-fired power plants are among the largest sources of air pollution in the United States. More than half a dozen separate Clean Air Act programs could possibly be used to control emissions, which makes compliance strategy potentially complicated for utilities and difficult for regulators. Because the cost of the most stringent available controls, for the entire industry, could range into the tens of billions of dollars, some power companies have fought hard and rather successfully to limit or delay regulations affecting them, particularly with respect to plants constructed before the Clean Air Act Amendments of 1970 were passed. The expected retirement of approximately 27 GW of coal-fired capacity by 2016 has been reported to the Energy Information Administration (EIA) by coal plant owners and operators, accounting for approximately 8.5% of U.S. coal-fired capacity. While the costs of compliance with new Environmental Protection Agency regulations are a factor, several other issues are cited by coal plant owners and operators as contributing to these retirement decisions including the age of coal-fired power plants, flat to modest electricity demand growth, the availability of previously underutilized natural gas combined-cycle power plants, and the lower price of natural gas due to shale gas development. Even coal plants which have made significant modifications to meet existing EPA regulations are being closed or mothballed due to a combination of low natural gas prices, and the inability to sell power into other markets. EIA expects coal to be a significant part of the U.S. power generation industry's future to well past 2030. But given price competition from natural gas, and emerging environmental regulations, that role will likely be smaller than in recent decades. Coal-fired generation is likely to face a challenging future.

In: U.S. and World Coal Production
Editor: Raul M. Schmidt
ISBN: 978-1-62618-976-8

Chapter 1

U.S. AND WORLD COAL PRODUCTION, FEDERAL TAXES, AND INCENTIVES*

Marc Humphries and Molly F. Sherlock

SUMMARY

Even though U.S. coal production remained strong over the past decade, reaching record levels of production, coal is losing its share of overall U.S. energy production primarily to natural gas. One of the big questions for the industry is how to penetrate the overseas market, particularly in steam coal, to compensate for declining domestic demand. As U.S. energy policy and environmental regulations are constantly debated, there is ongoing congressional interest in the role of coal in meeting U.S. and global energy needs.

The question may not be whether the domestic production of coal is here to stay but, rather, how much U.S. coal will be mined, what type, and under what regulatory framework.

Energy Information Administration (EIA) statistics show that more than half (55%) of U.S. coal reserves are located in the West, dominated by Montana and Wyoming, which account for 43%. When including the top five producing states (three of which are in the East), 70% of U.S. coal reserves are accounted for. The United States government owns about one third, or 87 billion short tons (BST), of U.S. domestic reserves.

* This is an edited, reformatted and augmented version of a Congressional Research Service publication, CRS Report for Congress R43011, from www.crs.gov, prepared for Members and Committees of Congress, dated March 14, 2013.

Coal production in the United States reached an all-time high of 1,174.8 million short tons in 2008, before declining to slightly under 1,100 million short tons from 2009 to 2011. Coal production on federal lands accounts for about 43% of U.S. production, according to the Bureau of Land Management (BLM). World coal production has increased by nearly 60% since 2002, most of the increase coming from China—up 130%.

Overall, U.S. coal production has been very strong over the past decade and if the industry is successful in penetrating the global market, primarily for steam coal, U.S. production may continue to grow faster than consumption. If recent trends continue, the U.S. coal industry will likely become more concentrated and produce more on federal lands. Businesses in the coal industry are subject to federal income taxes, but coal producers benefit from a number of federal tax provisions, commonly referred to as tax expenditures.

There are several congressional concerns related to coal production on federal lands. One concern is the potential for under-market-value coal auctions (sales), e.g., lease offers being accepted by the BLM with few competitive bids. In these cases, the federal government may not receive fair market value for the lease sale.

INTRODUCTION

Even though U.S. coal production remained strong over the past decade, coal is losing its share of overall U.S. energy production, primarily to natural gas. One of the big questions for the industry is how to penetrate the overseas coal market, particularly for steam coal, to compensate for declining domestic demand. In the past few years, coal production was at or near record levels and exports were up, while domestic demand declined. As U.S. energy policy and environmental regulations are constantly debated, there is ongoing congressional interest in the role of coal in meeting U.S. and global energy needs. The question may not be whether the domestic production of coal is here to stay but, rather, how much U.S. coal will be burned, what type, and under what regulatory framework.

The gap between U.S. coal production and consumption may continue to widen as low cost natural gas becomes more attractive to power plants (at least in the short run) and as older coal plants idle or close and uncertainties with emission regulations potentially inhibit new coal plant investments.

The Energy Information Administration (EIA) forecasts coal exports to continue to rise over their forecast period (2015-2040). Exports to the Asian market are expected to increase, but there are potential bottlenecks such as infrastructure (e.g., port development and transportation) that could slow export growth. Factors contributing to lower coal demand in the United States include low economic growth, high coal prices, lower natural gas prices, and coal plant retirements. *Table 1* illustrates the current coal supply-demand balance.

The coal industry is highly concentrated in the United States, with just a handful of major producers, operating primarily in four states, and the dominant holder of reserves (about one-third) being the U.S. government. Coal deposits (resources) are spread throughout the United States (see *Figure1*).

Table 1. U.S. Coal Supply and Demand (million short tons)

	2010	2011
Total Supply	1,117.0	1,120.0
- Production	1,084.0	1,096.0
- Imports	19.4	13.1
- Waste Coal Supplied	13.7	12.5
Total Demand	1,130.0	1,106.0
- Consumption	1,049.0	999.1
—Electric Power	975.1	928.6
- Exports	81.7	107.3
—Metallurgical	56.1	69.5
—Thermal	25.6	37.7
Stock Change	(13.0)	(7.1)
Unaccounted	0.2	20.7
Exports/Production	7.5%	9.8%

Source: U.S. Energy Information Administration (various reports), http://www.eia.gov/coal/data.cfm.

Notes: Units = million short tons (one short ton equals 2,000 pounds). May not sum due to rounding. Values in parenthesis are negative. According to EIA, waste coal is usable material that is a byproduct of previous coal processing operations, usually composed of mixed coal, soil, and rock (mine waste). Most waste coal is burned as-is in unconventional fluidized-bed combustors. For some uses, waste coal may be partially cleaned by removing some extraneous noncombustible constituents. Examples of waste coal include fine coal, coal obtained from a refuse bank or slurry dam, anthracite culm, bituminous gob, and lignite waste.

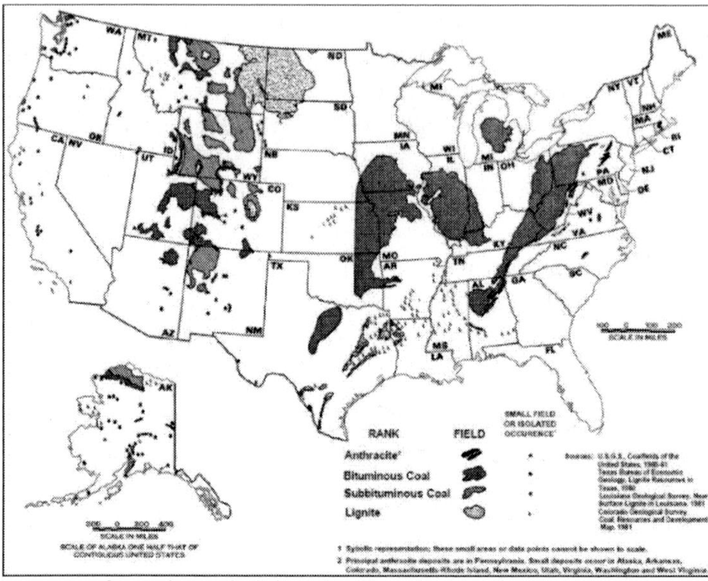

Source: Energy Information Administration, http://www.eia.gov/coal

Note: The Lower-48 United States is divided into three coal regions: Appalachia, Interior, and Western.

Figure 1. U.S. Coal Deposits.

There has been considerable congressional interest in coal, with over 100 bills that were introduced in the 112th Congress many of which addressed environmental and worker safety issues, neither of which are covered in this report. The discussion below is intended to provide some background and context for potential coal debates during the 113th Congress.

This report serves as a primer on U.S. and world coal resources and production and highlights some of the congressional interest related to coal production on U.S. federal lands. The report primarily describes the past 10 years of coal activity but also includes a discussion of future coal production projections and federal coal incentives, many of which do not expire.

Background Primer on Coal

Coal consists of the fossilized remains of ancient plant life that have been transformed through metamorphosis into carbon-rich mineral deposits.[1] It occurs as seams in sedimentary rock strata as old as 300 million years, though the most abundant deposits in the United States were deposited during the geologic Carboniferous period between 210 and 250 million years ago. Coal mineral classification considers type, rank, and grade. The plant life that coal originated from determines its type, and the degree of metamorphosis determines its rank, grade, and the amount of inorganic mineral matter present. Qualities such as moisture, carbon, sulfur, and ash content contribute to a coal's heating value as a fuel (measured in British thermal units [Btus]).

The content of sulfur is significant because of the sulfur dioxide (SO_2) emissions that occur during coal combustion. Under the Clean Air Act, there are federal limits on the amount of SO_2, among other pollutants, allowed from coal-fired power plants.[2] Western coal (low sulfur and low energy content) primarily produced in the Powder River Basin (PRB) of Wyoming,[3] is used generally for power generation, while eastern coal has been used domestically for power generation and exported for coking and metallurgical purposes. Moisture adds weight to the coal, increasing shipping costs while decreasing its heating value. Minerals deposited with the plants that formed coal create ash when coal burns. *Table 2* below illustrates the various coal classifications.

Table 2. Coal Classifications

	Type	Heating Value	Occurrence	
High	Anthracite (Hard Coal) Shiny black, High carbon content	15,000 Btu/lb. Typical use: metallurgy, domestic heating	Limited geographically to mainly Appalachia and Pennsylvania.	Low
→ RANK →	Bituminous (Soft Coal)	10,500 - 15,500 Btu/lb. Typical use: electric power generation, coke for steel making.	The most abundant U.S. coal type, found primarily in Appalachia and the Midwest.	← MOISTURE ←
	Sub-bituminous Dull-black	8,300 - 13,000 Btu/lb. Typical use: electric power generation, heating	Resources primarily found in Montana, Wyoming, Colorado, New Mexico, Washington, and Alaska. Most produced U.S. coal in 2011, accounting for 47% of production.	
Low	Lignite: Brownish-black High moisture High ash Low carbon Low heating value	4,000-8,300 Btu/lb. Typical use: electric power generation.	Resource mainly found in Texas, North Dakota, Louisiana, and Montana.	High

Source: International Energy Agency.

Coal Mining Methods

There are two primary mining techniques used in the United States: underground mining and surface mining. About 69% of U.S. coal comes from surface mines, while the remaining 31% comes from deep underground mines.

Underground Mining

There are two primary underground mining techniques: room and pillar (including conventional and continuous mining) and longwall. Room and pillar and continuous mining is practiced as follows: In flat-lying coal beds, conventional room and pillar techniques rely on cutting, drilling, blasting, loading, hauling, and roof bolting—a labor-intensive process. Long steel bolts properly spaced and driven into the roof are required. In continuous mining, the cutting, drilling, blasting, and loading are performed by a mechanical excavator known as a continuous miner. The continuous miner cuts and loads the coal. Together, room and pillar and continuous techniques account for 50% (conventional, room and pillar—2%; continuous—48%) of underground mining.[4]

The most efficient technique in underground mining is the longwall method, which employs a large machine with a rotating drum that moves back and forth across a wide coal seam. Once coal is removed by a longwall miner or other method, it is then moved out of the mine with conveyor belts or shuttle cars. About 50% of underground coal is produced using the longwall technique.

Surface Mining

Surface mining, also called "open-pit" or strip mining, entails blasting rock above the coal with explosives. This overburdened rock is then removed with huge electric shovels and draglines to reveal the coal seam. The coal seam in a surface mine is worked in long cuts by uncovering and removing coal then backfilling and reclaiming land in sequence. In other words, while coal extraction is taking place, as required by federal law, the reclamation work occurs in an adjacent area previously mined. Mountaintop removal mining is a form of strip mining.[5] Mountaintop removal mining generally removes a coal seam from one side of a mountain to the other. This is typically done in "steep-terrain" surface mining. Some of the overburden or "excess spoil" from the top of the mountain is placed in a valley fill. The placement of the excess spoil in valleys adjacent to mining areas remains controversial.[6]

The Coal Cycle

After being mined, coal goes through a cleaning prep facility, where it is cleaned and separated by grades. Cleaning upgrades the quality of the coal by removing some of the impurities such as rock, clay, and other ash-producing material. In the eastern United States, this refuse is generally pumped into an impoundment area often built near old underground mines in steeply sloping valleys. Once cleaned and separated, if necessary, the coal is stockpiled and shipped to the customer by rail, barge, truck, or conveyor. It usually takes more than one mode of transport for coal to reach its final destination. Utilities burn pulverized coal to produce high-pressure steam that powers an electric generator. As coal is burned, emissions are produced that contain sulfur dioxide, nitrogen oxides, carbon dioxide, particulate matter, ash, and mercury. A discussion of coal combustion emissions is found in other CRS reports.[7]

U.S. Coal Resources and Reserves

Assessing the amount of coal in the United States is quite complex. Two very different terms used to describe coal deposits—resources and reserves—are sometimes used interchangeably, but have very different meanings.

Resources provide a broad measure of coal production potential, or amount of coal "in the ground" summing the identified and undiscovered deposits of a minimum thickness.[8] Reserves are known quantities that can be produced given current prices and technology. Resource assessments are complex and rely on bed thickness as a major factor in determining recoverability, with direct correlation between bed thickness, depth, and recoverability. Coal rank (degree of metamorphism) and coal quality (heating value) contained in a unit of coal are factors as well. A February 2013 assessment by the U.S. Geological Survey (USGS) estimate the Powder River Basin coal resource alone at about 1.07 trillion short tons.[9]

A National Research Council (NRC) study cautions that it is not sufficient to know there is a vast resource base in the United States because only a small amount can be mined economically (the reserves) and advises caution when using coal resources to estimate the life of the U.S. coal base.[10] Resources, however, can become reserves. Reserve data is temporary and subject to change based on public policy, coal versus natural gas prices, production rates, transportation issues, exports, technology, and other conditions. Reserves are only an estimate of what might be recovered. Reserves must be developed from a better defined resource base according to the Virginia Center for Coal and Energy Research report.[11] A maturity index is used to measure the ratio of the amount of remaining economically recoverable resources to the previously mined tonnages. The Interior and Appalachian regions have less than a 0 index meaning their recoverable reserves are less than the amount already mined. Those areas are far more developed than the western regions. The remaining eastern resources may be thinner and of lower quality overall, which may lead to more mining underground than before because much of the surface resources have already been mined.[12]

The United States has the largest amount of coal reserves and resources in the world. The U.S. Energy Information Administration (EIA) estimates there are about 261 billion short tons of recoverable domestic coal reserves. The total demonstrated resource base (DRB) is estimated at about 484.5 billion tons.[13] EIA statistics show that more than half (55%) of U.S. coal reserves are located in the West, of which Montana and Wyoming together account for 43% (see *Table 3*). When including the top five producing states (three of which are in the East), 70% of U.S. coal reserves are accounted for. The United States government owns about one third, or 87 billion short tons (BST), of U.S. domestic reserves, followed by Great Northern Properties Limited Partnership (20 BST), and Peabody Energy Corporation (9 BST).[14] All together, the top three reserve owners account for about 45% of U.S. coal.

Table 3. U.S. Coal Reserves by State, 2011 (billion short tons)

State	Underground	Surface	Total	Percent of Total
Montana	35.9	38.8	74.7	28.8
Wyoming	22.9	14.9	37.8	14.6
Illinois	27.8	10.1	37.9	14.6
West Virginia	15.0	2.1	17.1	6.6
Kentucky	7.0	7.3	14.3	5.5
Pennsylvania	10.4	1.0	11.4	4.4
Ohio	7.6	3.7	11.3	4.4
Colorado	5.9	3.7	9.6	3.7
Texas	None	9.3	9.3	3.6
New Mexico	2.8	4.1	6.9	2.7
Others	12.8	15.6	28.4	10.7
Total	148.1	110.5	258.6	100.0

Source: EIA Annual Coal Report 2011.

Notes: Some differences from EIA's International Energy Outlook, 2011.

U.S. COAL PRODUCTION

In recent decades, the U.S. coal industry has changed significantly. Coal production has shifted from high-sulfur to low-sulfur driven by the steady demand from electric power plants (coal's primary customer) need to comply with environmental standards. Coal production has fluctuated since 2002 (see *Table 4*) but overall production has been higher than in previous decades. Coal production in the United States reached an all-time high (in tonnage) of 1,174.8 million short tons in 2008, before declining to slightly under 1,100 million short tons from 2009 to 2011. However, natural gas has been the recent fuel of choice for new power plants, reducing coal's domestic market share from 42% in 2011 to 32% by April 2012.

Table 4. U.S. Coal Production 2002-2011 (million short tons)

Year	Total	Eastern	Western	Underground	Surface
2002	1,094.3	492.9	601.4	357.4	736.9
2003	1,071.8	469.2	602.5	352.8	719.0
2004	1,112.1	484.8	627.3	367.6	744.5
2005	1,131.5	493.8	637.7	368.6	762.9
2006	1,162.7	490.8	672.0	359.0	803.7
2007	1,146.6	478.2	668.5	351.8	794.8
2008	1,171.8	493.3	678.5	357.1	814.7
2009	1,074.9	449.6	625.3	332.1	742.9
2010	1,084.4	446.2	638.2	337.2	747.2
2011	1,095.6	455.8	638.5	345.6	748.4

Source: Energy Information Administration, http://www.eia.gov/totalenergy/data/annual/pdf/sec7_7.pdf.

Out of the four major U.S. fuel sources—oil, natural gas, coal, and uranium—coal has the largest domestic reserve base and has accounted for the largest share of U.S. energy production in Btus since the early 1980s. In 2005, coal production was 33% of all U.S. energy production.[15] However, in 2011, natural gas surpassed coal and accounted for 30% of U.S. energy production because of sharp production increases, while coal fell to 28% of U.S. energy production. EIA's reference case (2011) predicts that coal will continue to lose energy production market share to natural gas as coal would drop from 28% to 24% by 2040 (see *Table 5*). Cases in which coal production rises more than the reference case involve assumptions of higher natural gas prices, lower coal prices, and higher economic growth. Coal production would grow at a 0.2% annual average rate through 2040 in the reference case.

Table 5. U.S. Energy Production in Btus (numbers are in percent of total)

	Coal	*Natural Gas*	*Crude Oil*	*Nuclear*	*Renewables*	*NGPLs*[a]	*Total*
1950	40	17	32	0	8	2	100
1960	25	30	35	negligible	7	3	100
1970	22	34	32	negligible	6	4	100
1980	28	30	27	4	8	3	100
1990	32	26	22	9	8	3	100
2000	32	28	17	11	9	4	100
2010	29	29	20	11	10	4	100
2025	25	32	21	10	11	NA	100
2040	24	34	17	10	14	NA	100

Source: U.S. EIA, Annual Energy Review 2011 (for historical data to year 2000). EIA Annual Energy Outlook 2013 Early Release Overview for data years 2010, 2025, and 2040.

a. NGPLs are natural gas plant liquids. EIA projections for 2025 and 2040. Percentages above may not add to 100 due to rounding.

Coal production on federal lands accounts for about 43% of U.S. production, according to the Bureau of Land Management (BLM). Coal production from the Powder River Basin (PRB), most of which is on federal lands, accounts for 41% of U.S. production, all of which is surfaced mined. Powder River Basin coal production is projected to increase, according to the EIA, becoming an even larger share of U.S. production as Interior coal has declined in recent years and is projected to grow slowly over the EIA forecast period (2016-2040). Appalachian coal has declined sharply over the past couple of decades and is projected to continue to decline through 2020 after which production is projected to increase, particularly for coking coal exports but with fewer shipments of steam coal for the U.S. market.[16]

Sub-bituminous surface mining in the western region (lower rank and lower sulfur coal) has become the dominant feature of U.S. coal mining since the mid-1970s. Surface mining has grown from 55% of coal production in 1975 to 69% of production in 2011. Western coal took the lead over eastern production in 1999 when it rose to 53% of production. Western production now accounts for 58% of U.S. coal production. U.S. coal production from the

West is projected to continue to dominate (reaching 68% of U.S. production) throughout the EIA forecast period (2016-2040).[17]

Eastern and underground coal may have peaked in 1990 at 630.2 and 424.5 million short tons, respectively.

A number of factors are behind this dramatic shift from underground eastern coal to western-based surface coal, and from bituminous coal to sub-bituminous coal including the ease of mining, coal utilization, mining conditions, mining technology, health and safety, environmental laws, and mining costs.[18] Five coal-producing states account for 72% of total U.S. coal output (see *Table 6*).

Table 6. U.S. Coal Production, Selected States, 2011 (million short tons)

State	Production	Percent of Total
Wyoming	438.7	40.0
West Virginia	134.6	12.3
Kentucky	108.8	9.9
Pennsylvania	59.2	5.4
Texas	45.9	4.2
Other	308.4	28.1
Total	1,095.6	100.0

Source: EIA /Annual Coal Report, 2011.

The National Research Council reported in 2007 that "the context for any assessment of future U.S. coal production is inextricably linked with the development of a national carbon emissions policy. Potential constraints on greenhouse gases (especially CO_2) emissions and the technical and economic feasibility of CO_2 control measures are the dominant issues affecting the outlook for the future of coal use over the next 25 years and beyond."[19] However, since 2007, there are several other factors that will influence U.S. coal production, such as, natural gas prices, the development of the infrastructure to support the use of natural gas for electricity, and how much and how quickly U.S. western steam coal can penetrate the international market.

Coal-Producing Industry

In 2011, the top five coal mining companies were responsible for about 58% of U.S. coal production, led by Peabody Energy Corp. with 18.5% and Arch Coal, Inc. with 14.6% (see *Table 7*). Other major producers include Alpha Natural Resources, Llc, Cloud Peak Energy, and CONSOL Energy, Inc. The concentration of production among the top five producers was similar in 2002 when they accounted for 46%. The two leading producers in 2002 were Peabody Energy Corp., with 13.7% of production, followed by Kennecott Energy Co. Inc. with 10.2% of production. The next three top producers included Arch Coal, Inc., RAG American Co., and CONSOL Energy, Inc.

Table 7. Leading U.S. Coal Producers

	2011		2002
Producer	Percent of Total	Producer	Percent of Total
Peabody Energy Corp.	18.5	Peabody Energy	13.7
Arch Coal, Inc.	14.6	Kennecott	10.2
Alpha Natural Resources	10.6	Arch Coal, Inc.	10.1
Cloud Peak	8.7	RAG American Co.	6.4
CONSOL Energy	5.7	CONSOL Energy	5.6

Source: EIA, Annual Coal Reports, 2002 and 2011.

The number of coal mining firms has decreased in the United States, while the size of the average mine and output per mine have increased. The number of mines has fluctuated over the past ten years, but generally has declined by 60% since the 1990s (e.g., from 3,430 in 1990 to 1,325 in 2011) as some of the smaller mines have become uneconomic.[20]

While coal production levels were at about the same level in 2011 as they were in 2002 and the number of coal mines declined, the number of coal miners has increased (75,466 to 91,611) over the same period. Production per miner hour has drifted downward since 2001 from 7.10 tons per hour to 5.55 tons per hour in 2011.[21]

The concentration of production at the global level is not as great. In 2010, the top five world coal producers account for about 18% of world production (Coal India (6%), Shenhua Group China (5%), Peabody Energy (2.8%), Datong Coal Mining Group-China (2.1%), and Arch Coal (2.1%). The top 30 coal firms produced 40% of world production (see *Table 8*).[22]

Table 8. Leading World Coal Producers, 2010

Producer	Percent of Total Production
Coal India	6
Shehua Group	5
Peabody Energy Corp.	2.8
Datong	2.1
Arch Coal, Inc.	2.1

Source: OECD/IEA.

COAL ON FEDERAL LANDS

There are substantial federal coal resources; about 957 billion short tons (excluding Alaska) according to a 2007 interagency report on an inventory of federal coal resources.[23] Production increases on federal land are possible given the industry's interest in seeking out international markets for PRB coal. The greatest growth opportunity and long term potential may be for thermal coal exports into the Asian markets.

The Leasing Process for Coal

The Bureau of Land Management (BLM) administers coal leasing on all federal lands. All BLM coal leasing is done competitively except in cases where a party holds a "prospecting permit" issued prior to the Federal Coal Leasing Amendments Act of 1976 or where contiguous acres are added to existing leases. The process for coal leasing on federal lands is similar to the process for oil and gas leasing. It is governed by Section 2 of the Mineral Leasing Act (MLA), as amended. Federal coal leasing is based on the BLM's Resource Management Plan (RMP) and the Forest Service (FS) Forest Plans. The BLM uses the following "planning screens" to focus on areas where there is the greatest interest or potential for leasing:[24]

1. Determine areas with coal potential;
2. Apply unsuitability criteria;
3. Apply multiple use conflict analysis; and
4. Consult with qualified surface owners (including private land owners).

There are two processes by which federal lands may be leased for coal production. The first is "regional coal leasing," in which BLM selects tracts for leasing as needed to meet regional requirements as outlined by "regional coal teams" composed of BLM officials and interested state and local parties. The second is leasing on application whereby mining companies submit an application to lease certain tracts.[25]

About 70% of the estimated federal coal resources on 3.6 million acres have not been screened under the BLM's land use planning process because of low industry interest.[26]

Regional Coal Leasing

The Secretary of the Interior establishes regional coal leasing levels after receiving input from the states and other relevant stakeholders. Areas are then delineated and ranked according to interest for coal leasing. After ranking and selection, BLM produces a regional coal leasing environmental impact statement (EIS) to satisfy the requirements of the National Environmental Policy Act (NEPA).

After further consulting at the local, state, and federal levels, the BLM adopts a lease sale schedule. The BLM establishes fair market value (FMV) and maximum economic recovery for the tracts before the lease sale begins.[27] BLM does not accept any bid for less than the FMV as determined. Currently, the minimum bid is established in regulation (43 C.F.R. 3422) at no less than $100 per acre.

After notice, the BLM conducts the lease sale and awards the lease to the highest bidder that meets the lease sale requirements. BLM reserves the right to reject any and all bids for any reason.[28]

Leasing on Application

Under this process, a coal producer submits an application to the BLM indicating an interest in certain tracts that are included in its coal land use plan. The application must contain certain data intended to assist BLM in conducting the environmental analysis needed to satisfy the requirements of National Environmental Policy Act (NEPA).

Then, the BLM determines the FMV and maximum economic recovery for the proposed lease tract, and consults with the same parties with whom consultation is required for regional leasing as described above. After these requirements are met, the lease sale is conducted in the same manner applicable to regional coal lease sales.[29]

Lease Terms and Conditions for Coal

General Statutory Restrictions

Under U.S. and state laws, only U.S. citizens, associations of U.S. citizens, and corporations organized under U.S. laws may bid on and lease coal on federal land. No entity is permitted to own or control coal leases with an aggregate acreage in excess of 75,000 acres in any one state or more than 150,000 acres in the United States.[30]

Coal Lease Terms

BLM-issued coal leases are for initial terms of 20 years, with automatic extension "for so long thereafter as coal is produced annually in commercial quantities from that lease." In addition to rental payments of not less than $3 per acre, lessees are required to make payment to the government of a royalty of at least 12.5% in amount or value of coal that is recovered from leased land that is surfaced mined. The federal rate for underground coal mining is 8%. All leases are subject to the condition of diligent development and continued operation.[31] Lessees must also furnish bonds sufficient to ensure compliance with the terms and conditions of the lease.[32]

Coal Revenues (Federal and State)

Historically, federal coal production has generated significant revenues for the federal government and the states in which it is produced. States receive 50% of revenues generated in their state (minus administrative costs of 2%). Total revenues have grown rapidly since the 1990s, reflecting higher coal values and volumes. For instance, royalty revenue generated from federal coal leases doubled from 1983 to 1986, then doubled again by 1990, reaching $236 million. Revenue continued to rise, reaching $434 million in 2002 and $774.1 million in 2011. The states' share in 2011 was $387.2 million, with Wyoming receiving $319.6 million or 82% of the total disbursed to the states. Revenues from federal coal leases reached an all-time high in 2009 at $780.4 million, providing states with revenues of $348.6 million. Wyoming has been the dominant coal producer on federal lands, receiving the majority of federal coal revenues disbursed to the states.[33]

Federal Coal Resources

Under Section 437 of the Energy Policy Act of 2005, Congress directed the Administration to conduct an inventory and assessment of federal coal resources and restrictions to their development. The study was to identify lands available for coal development and restrictions on their land and to identify environmentally compliant and super-compliant resources (based on sulfur dioxide emissions per million Btus).[34] The study evaluated the Powder River Basin (PRB) federal land and coal resources.[35]

The PRB contains 550 BST or 58% of the federal resources assessed, and PRB represents 88% of coal produced on federal lands. [36] The total federal estate surveyed was 5.4 million acres (including split estate lands whereby the surface owner and mineral rights owner are two separate parties).

There are three major categories of land classification in the study:

1. leasing available under standard lease terms or with no surface occupancy;
2. leasing permitted with restrictions (possible leasing); and
3. leasing prohibited.

Leasing is prohibited on 591,000 acres in the PRB, which is estimated to contain 5.9% of the potential federal coal. Leasing is "possible" on 4.3 million acres containing an estimated 84.3% of federal coal resources. Leasing is available under standard lease terms on 82,000 acres and available with no surface occupancy on 431,000 acres, containing 5% and 4.3 % of federal coal resources, respectively. See *Figure 2.*

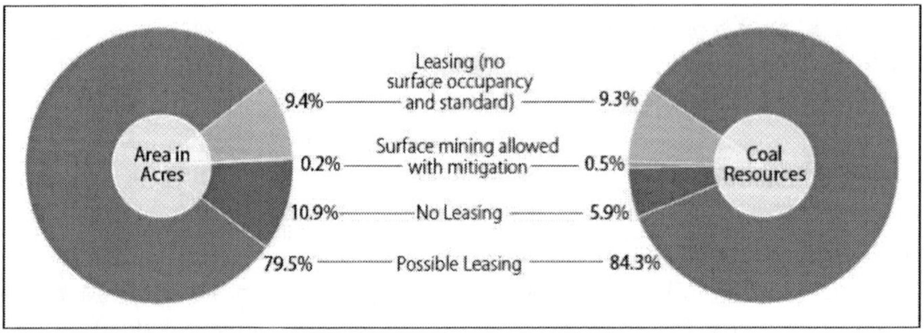

Source: U.S. Departments of Energy, Interior, and Agriculture.

Figure 2. PRB Federal Land and Coal Resources by Access Category.

Congressional Concerns

There are several congressional concerns related to coal production on federal lands. A key concern raised by stakeholders is the potential for under-market-value coal auctions (sales), e.g., lease offers being accepted by the BLM with few competitive bids and the federal government possibly not receiving fair market value for the lease sale.

The BLM has established a floor bid price of $100 per acre but typically bids are offered and accepted on a dollar-per-ton basis. The minimum acceptable bid amount is established by BLM and kept confidential. Most recently, these bids have ranged from $0.25 per ton to over $1.00 per ton, based on the quality of the reserves, among other factors. Coal reserve estimates are established internally by the BLM and often based on exploration information gathered by the coal industry. The $100 per acre floor price or the minimum dollar bids per ton are added to the potential royalty revenue to represent the fair market value of the tracts being leased.

Other questions arise regarding whether the federal government is receiving the fair market value based on the value of the coal, particularly in the international market, which would include examining arms-length and non-arms-length transactions as well as price, volume, and the royalty rate.[37] For arms-length sales, the Office of Natural Resources Revenue (ONRR) relies on the "gross proceeds accruing to the lessee" (minus allowable deductions, such as transportation and washing) to value coal for royalty purposes. For non-arms-length transactions, coal valuation regulations (30 C.F.R. 1206.257 (c) (1)(2)) provide ONRR guidance on valuing coal sales.[38] In a joint letter (January 3, 2013) to Secretary Ken Salazar, Senators Wyden and Murkowski outlined the possibility of underpayment of federal royalties and raised specific questions for more information about the royalty management program.[39]

Another associated issue of concern is the opportunity for coal producers to lease contiguous tracts non-competitively. The BLM justifies this practice on the grounds that there would be little interest by others and that the current practice makes the coal more likely to be developed. But to critics, this practice offers an unfair advantage to the existing coal producer.

The PRB area was declared a non-coal producing region by the BLM in 1990 because of the low expression of leasing interest at that time. For example, the BLM reports that industry interest fell from 46 expressions of interest in 1982 to six expressions of interest in 1988.[40] This designation allows for lease-by-application to occur and is of concern to many because it offers fewer opportunities for public input on a potential lease sale, according to the BLM. Environmental groups are arguing for the PRB region to be recertified as a coal producing area. The Powder River Regional Coal Team (RCT) reviews the region's status and could recommend to the BLM Director that the region be recertified. The RCT consists of representatives from BLM, state and county agencies, and Native American tribes in Wyoming and Montana.

TAX ISSUES[41]

Currently, there are a number of federal tax incentives available that support coal production, clean coal or advanced coal technologies, and coal mine safety. Certain states impose severance taxes—an excise tax on natural resource extraction—on coal.

Federal Tax Payments

Businesses in the coal industry are subject to federal income taxes. In 2009, 1,150 coal mining corporations filed tax returns with the Internal Revenue Service (IRS), paying a total of $207.4 million in corporate income taxes.[42] In addition to corporate income taxes, taxes were paid by coal companies structured as partnerships or other forms of pass-through entities on individual income tax returns.[43] Domestically mined coal is also subject to an excise tax, with excise tax revenues dedicated to the Black Lung Disability Trust Fund. The tax rate is $1.10 per ton of coal produced from underground mines and $0.55 per ton of coal produced from surface mines, subject to a maximum tax rate of 4.4% of the coal's sales price. In 2010, black lung excise tax collections were $610.1 million.[44] In FY2011, the balance of the Black Lung Disability Trust Fund (BLDTF) was negative $5.5 billion.[45] The black lung excise tax is set to be collected until the Trust Fund has repaid all amounts borrowed from the general fund, or until December 31, 2018.

Federal Tax Incentives

Coal producers benefit from a number of federal tax provisions, commonly referred to as tax expenditures (see *Table 9).*[46] Some of these incentives support conventional coal mining operations—specifically, the ability to expense (i.e., deduct immediately) exploration and development costs and the ability to claim percentage as opposed to cost depletion. Other incentives, such as the production tax credit (PTC) for refined coal and the manufacturers.[47] Mining is considered a qualified manufacturing activity for the purposes of claiming this deduction. Businesses that earn income through coal mining activities may also be allowed to structure as master limited partnerships (MLPs), a type of business structure that is associated with certain tax benefits.[48] The Section 199 deduction and the ability to structure as an MLP, while not unique to the coal industry, serve to reduce taxes paid by coal-related businesses. The President's FY2013 Budget proposes eliminating various tax incentives for coal as part of a broader strategy to "phase out subsidies for fossil fuels."[49] Specifically, the President's proposal would repeal expensing of exploration and development costs, require that coal mines use cost depletion rather than percentage depletion, and repeal the capital gains treatment of coal royalties. Eliminating these three incentives would generate an estimated $2.6 billion in additional federal revenues over the FY2013 to FY2022 budget window.[50] The President's FY2013 Budget also proposes eliminating the Section 199 production activities deduction for coal, using the revenues generated to increase the Section 199 production activities deduction for certain advanced manufacturing activities.[51]

Severance Taxes

Severance taxes are taxes imposed on the removal of natural resources from land or water. Severance taxes are generally charges based on the value or volume of the natural resource being extracted, and are thus independent of income or profits. Several states impose severance taxes on coal.[52] Severance taxes on coal are not collected at the federal level.

World Coal

World Coal Reserves and Resources

Table 10 illustrates global ranking of coal reserves. The United States has about 261 BST in reserves (27.4% of world total) followed by Russia with an estimated reserve base of 173 BST (18.3%). China is third, with 126 BST (13.3%). Taken together, the top three countries hold 59% of the world's recoverable coal reserves. When India and Australia are added, the top five coal producing countries hold 75% of world recoverable coal reserves. This credit for investment in clean coal facilities are designed to support c. advanced coal technologies.

Table 9. Coal-Related Federal Tax Incentives

Incentive	Description	Expiration
Expensing of Exploration and Development Costs	Taxpayers can expense (deduct immediately) coal mine exploration expenditures, with deducted amounts recaptured once the mine is producing or is sold. C corporations can only expense 70% of qualifying costs, with the remaining 30% recovered over 5 years. Alternatively, mine exploration expenditures can be amortized over a 10-year period.	Does Not Expire
Percentage Depletion	Capital costs can be recovered using percentage depletion. For coal, depletion deductions equal to 10% of gross income from the property are allowed. The deduction is limited to 50% of taxable income from the property.	Does Not Expire
Coal Royalties Treated as Capital Gains Income	Dispositions of coal royalties are treated as long-term capital gains, which are taxed at reduced rates.	Does Not Expire
Credit for Investment in Clean Coal Facilities Exclusion of Special Benefit for Disabled Coal Miners	A 30% tax credit is available for integrated gasification combined cycle (IGCC) and advanced coal-based electricity generation technologies (ACBGT) projects. Tax credits are allocated to qualifying projects jointly by the Treasury and Department of Energy (DOE). Disability payments to former coal miners out of the Black Lung Disability Trust Fund are excluded from taxable income.	Allocation Based – Limited Funds Available Does Not Expire

Source: CRS and the Internal Revenue Code (IRC).

Notes: Generally available tax incentives from which the coal industry benefits, such as the Section 199 production activities deduction, are not included in this table.

There are also tax provisions designed to encourage investment in mine safety equipment and mine rescue training. These two incentives were initially enacted on a temporary basis, but have been extended in the past as part of "tax extenders" legislation. In addition, the coal industry benefits from a number of tax incentives that are available to multiple industries. For example, the Section 199 production activities deduction is available to all domestic indicates a great deal of coal reserves concentration by country. The U.S. government alone would rank fourth (9%) behind China in reserve holdings.

Table 10. World Recoverable Coal Reserves by Rank, 2009 (billion short tons)

Country	Bituminous and Anthracite	Sub- bituminous	Lignite	Total	Percent of Total
United States	119.2	108.2	33.2	260.6	27.4
Russia	54.1	107.4	11.5	173.1	18.3
China	68.6	37.1	20.5	126.2	13.3
Other Non-OECD	42.2	19.1	40.1	101.4	10.7
Europe and Eurasia					
Australia and New Zealand	40.9	2.5	41.4	84.8	8.9
India	61.8	0.0	5.0	66.8	7.0
Africa	34.7	0.2	0.0	34.9	3.7
Other	101.1	31.8	103.7	236.5	10.7
World Total	445.7	287.0	215.3	948.0	100.0

Source: IEO, 2011.

World Coal Production

World coal production has increased by nearly 60% since 2002, with most of the increase coming from China—up 130%. China accounted for about 50% of coal production in 2011, up from 34% in 2002. The data illustrate that other countries such as Columbia, Indonesia, and India also had significant production increases since 2002. India's coal production grew by 60% over the past 10 years, while Indonesia's production more than tripled. Australia increased coal production by 26% over the same time period. Indonesia, a major coal exporter, is likely to use a larger share of its production for domestic consumption. See *Figure 3.*

Projected World Coal Production

EIA and OECD Outlook

EIA projects world coal output to rise by 40% over its most recent forecast period (2011-2035), but only by 11% over the next 10 years.[53]

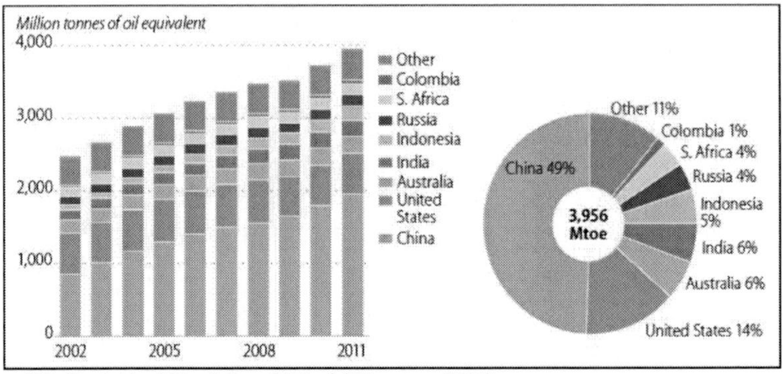

Source: BP Statistical Review of World Energy, 2012.

Figure 3. World Coal Production by Percent.(2011).

China would account for 61% of coal production increases through 2035. China accounts for 47%-50% of production now and is expected to account for about 51% in 2035, according to EIA projections. India would increase its production by 50% over the forecast period, as non-OECD Asian (including Indonesian) coal production is projected to grow by about 50%. EIA projects U.S. production to rise by 17% over the forecast period and coal production capacity to be large enough to satisfy the U.S. coal market. The EIA projects overall U.S. coal exports, measured in Btus, to rise by over 80% from 2009-2035, most of that increase coming from coking coal exports, which more than double over the forecast period. Steam coal exports would rise by 37%.

The OECD/IEA World Energy Outlook report (2011) makes projections based on a "New Policies Scenario."[54] According to OECD "new policies" projections, global coal output would rise 18% between 2009 and 2020, then remain essentially flat through 2035.[55] Growth in coal production is mostly in China and in other non-OECD countries, such as India and Indonesia. India would increase production by 70% and also increase its share of world output from 5.6% to over 10%. Under the New Policies scenario, U.S. coal output would decline from 14.1% to 12% of world coal output over the forecast period.

Overall, coal output from OECD countries would fall by 14%, while coal output from non-OECD countries would rise by 32%, according to the report. The OECD analysis shows significant gains for U.S. coal supply onto the world market—exports growing by 65% between 2009 and 2035 (most of which would be coking coal)—and projects a weak outlook for coal in the domestic market, which is considered by OECD to be the primary driver of U.S. coal output.

CONCLUSION

Global coal production and demand has grown significantly over the past decade and coal will likely be the dominant fuel for power needs over the EIA forecast period, as Asian and many emerging economies grow and more rapidly increase their demand for electric power.

Despite the regulatory concerns, U.S. coal production was robust over the last decade, reaching an all-time production high (in tonnage) in 2008, before a slight decline following the great recession. U.S. coal production is forecast to rise at a rate (<1%) slightly above projected U.S. demand through 2040, according to the EIA. In 2011, coal fell below natural gas as the leading source of primary energy production in the United States and will unlikely reclaim the top spot that it held for nearly the past three decades.

If trends continue, the U.S. coal industry will likely become more concentrated and produce more on federal lands. This may raise the issue that lease sales could become no more competitive in the future than they are today and possibly even less competitive. But since fair market value returns to the government consider all of the revenue streams, not just bonus bids at the lease sale, it will become increasingly important for the federal government to capture the correct valuation of federal coal (price x volume) and assign the appropriate royalty rate. Most of the federal coal (94%) in the PRB is accessible or possibly accessible (pending land use plans or surface owner consent), but 70% of the resource base has not been screened for leasing because of low interest and higher costs. This lack of interest could place a limit on just how much federal coal actually gets developed.

Overall, U.S. coal production has been very strong over the past decade and if the industry is successful in penetrating the global market, primarily for steam coal, U.S. production may continue to grow faster than consumption. Although U.S. metallurgical coal will likely continue to be in demand on the world market, higher than predicted steam coal exports could further widen the gap between U.S. coal production and consumption.

End Notes

[1] In geology, metamorphism is the solid state change in the structure, texture, or composition of rocks caused by heat, pressure, or stress.

[2] There are also state limits in some cases.

[3] PRB coal is located primarily in Montana and Wyoming.

[4] Meeting Projected Coal Production Demands in the USA, Upstream Issues, Challenges, and Strategies, prepared by the Virginia Center for Coal and Energy Research, Virginia Polytechnic Institute and State University, 2008, pp. 80-82.

[5] For an in-depth discussion of mountaintop removal mining see CRS Report RS21421, Mountaintop Mining: Background on Current Controversies, by Claudia Copeland.

[6] Meeting Projected Coal Production Demands in the USA, Upstream Issues, Challenges, and Strategies, p. 75.

[7] See CRS Report R42950, Prospects for Coal in Electric Power and Industry, by Richard J. Campbell, Peter Folger, and Phillip Brown, and CRS Report R41914, EPA's Regulation of Coal-Fired Power: Is a "Train Wreck" Coming?, by James E. McCarthy and Claudia Copeland.

[8] U.S. Geological Survey, The National Coal Resource Assessment Overview, USGS Professional Paper 1625-F, Editors Brenda S. Pierce and Kristin O. Dennen, Chapter D, p. 1, 2009.

[9] USGS, Assessment of Coal Geology, Resources, and Reserve Base in the Powder River Basin, Wyoming and Montana, Fact Sheet 2012-3143, February 2013.

[10] National Research Council, Committee on Coal Research, Technology, and Resource Assessments to Inform Energy Policy, Coal: Research and Development to Support National Energy Policy, National Academies Press, 2007.

[11] Meeting Projected Coal Production Demands in the USA, Upstream Issues, Challenges, and Strategies, prepared by the Virginia Center for Coal and Energy Research, Virginia Polytechnic Institute and State University, 2008, pp. 80-82.

[12] Ibid.

[13] Inventory of Assessed Federal Coal Resources, August 2007. (42% of the DRB is located in the PRB and an estimated 203.5 billion short tons in the PRB is on federal lands.) The demonstrated resource base is defined by the USGS as measured and indicated reserves plus sub-economic resources.

[14] National Mining Association, 2010 Coal Producers Survey, May 2011.

[15] EIA, Annual Energy Outlook (AEO), Early Release Overview, 2012, p. 12.

[16] AEO, 2012, p. 99.

[17] EIA , Annual Energy Outlook 2013, Early Release Overview.

[18] Meeting Coal Production Demands in the U.S.A, Upstream Issues, Challenges, and Strategies.

[19] National Research Council, Committee on Coal Research, Technology, and Resource Assessments to Inform Energy Policy, Coal: Research and Development to Support National Energy Policy, National Academies Press, 2007.

[20] EIA, Annual Coal Report, 2011.

[21] National Mining Association, Most Requested Statistics, U.S. Coal Industry, 2012.

[22] OECD/IEA, 2011, p. 418-419.

[23] Inventory of Assessed Federal Coal Resources and Restrictions to Their Development, In compliance with the Energy Policy Act of 2005, P.L. 109-58 section 437, prepared by U.S. Departments of Energy, Interior, and Agriculture.

[24] Ibid, p.12.

[25] 43 C.F.R. 3420 Coal Management, Competitive Leasing.

[26] Inventory of Assessed Federal Coal, p. xi.

[27] BLM, Coal Resources Frequently Asked Questions, http://www.blm.gov/energy/coal resources.

[28] 43 C.F.R. 3420 Coal Management, Competitive Leasing.

[29] 43CFR 3425, Leasing on Application.

[30] BLM at http://www.blm.gov/energy/coal.

[31] Diligent development is a statutory requirement that the lessee begin producing commercial quantities of coal within the 10-year primary lease term. If not, the lease is terminated.

[32] Ibid.

[33] Office of Natural Resources Revenue website at http://www.onrr.gov.

[34] "Lands available" are areas in a Resource Management Plan that are leased or could be leased for coal development.

[35] Inventory of Assessed Federal Coal Resources and Restrictions to Their Development, prepared by U.S. Departments of Energy, Interior, and Agriculture in compliance with the Energy Policy Act of 2005, P.L. 109-58, August 2007.

[36] A broad category of measurement that would include the demonstrated reserve base and economically recoverable reserve (as discussed in the Resource and Reserve section of this report), but using a broader resource assessment methodology that includes the following categories: hypothetical, inferred, indicated, and measured.

[37] An arms-length contract is one in which the contract is between independent, non-affiliated parties, and those parties must have opposing economic interest. Affiliations could be evaluated by looking at the corporate ownership. A nonarms-length contract is one between affiliates.

[38] The regulations discuss five sequenced benchmarks that may be used to determine value: gross proceeds for a comparable arms-length sale; use of coal price reported to the Public Utility Commission by the utility; price of delivered coal as reported to the EIA; other relevant matters such as the spot market price paid by unaffiliated producers with comparable contracts; or the net back valuation method (which nets out certain downstream costs to get to a free-on-board mine price). There are a number of conditions that determine which benchmark is to be used.

[39] http://www.wyden.senate.gov/news/press-releases/wyden-murkowski-seek-answers-on-coal-royalty-payments

[40] BLM Environmental Assessment of the North Antelope and Rochelle Coal Lease Applications for a Powder River Coal Company.

[41] Authored by Molly Sherlock with CRS's Government Finance and Taxation section.

[42] Internal Revenue Service, Statistics of Income—2009: Corporate Income Tax Returns, Washington, DC, http://www.irs.gov/pub/irs-soi/09coccr.pdf.

[43] The IRS Statistics of Income (SOI) data does not provide information on coal mining partnerships and S corporations. In 2009, there were 1,158 partnerships engaged in other mining activities, which include coal, metal ore, and nonmetallic mineral mining. These partnerships reported $1.0 billion in net income in 2009. This income flowed through and was reported on individual income tax returns. Data on income taxes paid by coal partnerships is not available.

[44] See IRS SOI Excise Tax Statistics, Table 20, Federal Excise Taxes Reported to or Collected by the Internal Revenue Service, Alcohol and Tobacco Tax and Trade Bureau, and Customs Service, by Type of Excise Tax, Fiscal Years 1999- 2010, available at http://www.irs.gov /taxstats/bustaxstats/article/0,,id=97148,00.html.

[45] Receipts collected for the fund have not been sufficient to cover compensation payments and medical expenses for former coal mine employees. Over time, tax revenues have been supplemented with appropriations from the general fund and the issuance of debt.

[46] Background information on specific tax provisions can be found in U.S. Congress, Senate Committee on the Budget, Tax Expenditures: Compendium of Background Material on

Individual Provisions, committee print, prepared by Congressional Research Service, 111th Cong., December 2010, S.Rept. 111-58.

[47] For additional background, see CRS Report R41988, The Section 199 Production Activities Deduction: Background and Analysis, by Molly F. Sherlock.

[48] For background on the MLP structure, see CRS Report R41893, Master Limited Partnerships: A Policy Option for the Renewable Energy Industry, by Molly F. Sherlock and Mark P. Keightley.

[49] Department of the Treasury, General Explanations of the Administration's Fiscal Year 2013 Revenue Proposals, Washington, DC, February 2012, p. 120.

[50] Ibid., p. 204.

[51] See CRS Report R41988, The Section 199 Production Activities Deduction: Background and Analysis, by Molly F. Sherlock.

[52] Information on state-level severance taxes can be found on the National Conference of State Legislatures website, at http://www.ncsl.org/issues-research/budget/2011-state-severance-tax-collections.aspx. Severance taxes are distinct from state-level income taxes, which may also be imposed on coal-related businesses.

[53] Output is measured in quadrillion BTUs.

[54] OECD's New Policy Scenario assumes "cautious implementation of the policy commitments and plans that have been announced by countries around the world" to reduce carbon emissions.

[55] Output is measured in million tons of coal equivalents.

In: U.S. and World Coal Production
Editor: Raul M. Schmidt
ISBN: 978-1-62618-976-8

Chapter 2

MOUNTAINTOP MINING: BACKGROUND ON CURRENT CONTROVERSIES*

Claudia Copeland

SUMMARY

Mountaintop removal mining involves removing the top of a mountain in order to recover the coal seams contained there. This practice occurs in six Appalachian states (Kentucky, West Virginia, Virginia, Tennessee, Pennsylvania, and Ohio). It creates an immense quantity of excess spoil (dirt and rock that previously composed the mountaintop), which is typically placed in valley fills on the sides of the former mountains, burying streams that flow through the valleys. Mountaintop mining is regulated under several laws, including the Clean Water Act (CWA) and the Surface Mining Control and Reclamation Act (SMCRA).

Critics say that, as a result of valley fills from mountaintop mining, stream water quality and the aquatic and wildlife habitat that streams support are destroyed by tons of rocks and dirt. The mining industry argues that mountaintop mining is essential to conducting surface coal mining in the Appalachian region and that it would not be economically feasible there if operators were barred from using valleys for the disposal of mining overburden. Critics have used litigation to challenge the practice. In a number of cases discussed in this report, environmental

* This is an edited, reformatted and augmented version of a Congressional Research Service publication, CRS Report for Congress RS21421, from www.crs.gov, prepared for Members and Committees of Congress, dated August 1, 2012.

groups have been successful at the federal district court level in challenging issuance of permits for mountaintop mining projects, but each has been later overturned on appeal. Nonetheless, the criticisms also have prompted some regulatory changes, also discussed here.

In June 2009, officials of the Environmental Protection Agency (EPA), the U.S. Army Corps of Engineers (Corps), and the Department of the Interior signed a Memorandum of Understanding outlining a series of administrative actions under these laws to reduce the harmful environmental impacts of mountaintop mining and surface coal mining in Appalachia. The plan includes a series of near-term and longer-term actions that emphasize specific steps, improved coordination, and greater transparency of decisions. The actions are being implemented through regulatory proposals, guidance documents, and review of applications for permits to authorize surface coal mining operations in Appalachia. Viewed broadly, the Administration's combined actions on mountaintop mining displease both industry and environmental advocates. The additional scrutiny of permits and more stringent requirements have angered the coal industry and many of its supporters. Controversy also was generated by EPA's January 2011 veto of a CWA permit that had been issued by the Corps for a surface coal mining project in West Virginia. At the same time, while environmental groups support EPA's steps to restrict the practice, many favor tougher requirements or even total rejection of mountaintop mining in Appalachia. Federal courts have recently rejected several of the Administration's actions, including overturning EPA's veto of the West Virginia mine permit, enhanced permit review procedures, and EPA guidance on factors used in evaluating water quality impacts of Appalachian surface mining permits.

This report provides background on regulatory requirements, controversies and legal challenges to mountaintop mining, and recent Administration actions. Congressional interest in these issues also is discussed, including legislation in the 111th Congress seeking to restrict the practice of mountaintop mining and other legislation intended to block the Obama Administration's regulatory actions. Attention to EPA's veto of the West Virginia mining permit and other federal agency actions has increased in the 112th Congress. Several bills have been introduced to clarify or restrict EPA's authority to veto CWA permits issued by the Corps (H.R. 457/S. 272; H.R. 517; H.R. 960/S. 468; and H.R. 2018, which the House passed in July 2011).

WHAT IS MOUNTAINTOP MINING?

The environmental, economic, and societal impacts of the surface mining practice termed mountaintop removal mining have attracted considerable attention. This type of surface mining occurs in an area of approximately 12 million acres located in portions of Kentucky, West Virginia, Virginia, Tennessee, Pennsylvania, and Ohio.

As its name suggests, mountaintop removal mining involves removing the top of a mountain in order to recover the coal seams contained in the mountain. Explosives are used to break the mountain's rock, and massive earth-moving equipment, often including equipment called draglines, removes the spoil (i.e., the dirt and rock that composed the mountaintop over or between the coal seams). While federal law calls for excess spoil to be placed back in the mined areas—returning the lands to their approximate original contour (AOC)—that result ordinarily cannot be accomplished with mountaintop mining because broken rock takes up more volume than did the rock prior to mining and because there are stability concerns with the spoil pile. Mountaintop removal creates an immense quantity of excess spoil, which is typically placed in valley fills on the sides of the former mountains. One consequence is that streams flowing through the valleys are buried.

All types of surface and underground coal mining in Appalachia generate excess spoil fills due to the increased volume of broken rock, limitations on the steepness and height to which broken rock may be placed to achieve a stable slope, and the steep topography of the region. Large mines may be surrounded by several valley fills. Depending on the local topography and the profile of those valleys, a single fill may be over 1,000 feet wide and over a mile long.

While mountaintop removal mining has been practiced in some form since the 1960s, it became a prevalent coal mining technique in parts of central Appalachia during the 1990s for several reasons. First, as the demand for electricity increased, so has the demand for the relatively clean-burning, low-sulfur coal found in Appalachia. Second, coal supplies near the surface have been significantly depleted. Third is the development of large surface mining equipment (draglines) capable of moving over 100 cubic yards of earth in a single scoop.

For many years, excess spoil from coal mining was generally placed in the extreme headwaters of streams, affecting primarily ephemeral streams that flow intermittently only in direct response to precipitation in the immediate watershed. Because smaller upstream disposal sites are exhausted and because of the increase in mountaintop mining activity, today the volume of a single stream fill can be as much as 250 million cubic yards. As a result, streams are eliminated, stream chemistry is harmed by pollutants in the mining overburden, and downstream aquatic life is impaired. EPA estimates that since

1992 almost 1,200 miles of Appalachian streams were buried by surface coal mining practices. The cumulative effects of such surface coal mining operations include deforestation, which has been linked to harm in aquatic communities; accelerated sediment and nutrient transport; and increased algal production, as well as possible human health impacts.[1]

Regulatory Setting

Regulation of valley fills associated with mountaintop removal mining is primarily under the authority of two federal statutes, the Surface Mining Control and Reclamation Act (SMCRA, 30 U.S.C. §1201) and the Clean Water Act (CWA, 33 U.S.C. §1252), and involves several federal and state agencies. The two laws provide for separate regulatory programs with different purposes and different permitting requirements and procedures. For example, the CWA focuses primarily on regulating discharges into waters of the United States, while SMCRA regulates a broad range of environmental and other impacts of surface coal mining and reclamation operations.

SMCRA addresses the necessary approvals for surface mining operations, as well as inspection and enforcement of mine sites until reclamation responsibilities are completed and all performance bonds are released. SMCRA permits may be issued by the Office of Surface Mining, Reclamation and Enforcement (OSM), U.S. Department of the Interior, or by qualified states, only if it has been shown that the proposed mining activities will satisfy general performance standards applicable to all surface coal mining operations.[2] Among those standards, SMCRA addresses disturbances at the mine site and in associated offsite areas, as well as the quality and quantity of water in surface and ground water systems both during and after surface coal mining operations. While SMCRA generally requires that surface-mined areas be reclaimed so that they closely resemble the general surface configuration of the land prior to mining (i.e., AOC), the law and OSM regulations allow a variance from AOC for mountaintop mining operations under certain conditions.

The CWA prohibits the discharge of any pollutant from any point source into the waters of the United States, except in compliance with a permit issued under one of the two permit programs established by the statute. The two permit programs are the National Pollutant Discharge Elimination System (NPDES) program, administered by the Environmental Protection Agency (EPA) under CWA Section 402, and the dredge and fill permit program administered by the U.S. Army Corps of Engineers (Corps) under CWA Section 404.[3] Mountaintop mining and other surface coal mining operations typically require both types of permits—a Section 404 permit for the discharge of mining overburden into waters of the United States, and a Section 402 permit for discharges from sediment ponds, on-site coal preparation facilities, and stormwater discharges from the mine site. The two permit programs employ different regulatory approaches and criteria.

The NPDES program focuses primarily (but not exclusively) on discharges from industrial operations and sewage treatment plants. Section 402 permits must include limitations on the quantities, rates, and concentrations of pollutants, called effluent limitations, that reflect treatment with available pollution control technology and any more stringent limitations necessary to meet state-established water quality standards for the receiving water. The standard for issuance of a Section 402 permit is compliance with pollutant limitation and control provisions in the CWA.

The Corps and EPA have complementary roles in implementing the Section 404 permit program. Under Section 404, the Corps issues permits for the discharge of dredged or fill material, using a set of environmental guidelines promulgated by EPA in conjunction with the Corps. These guidelines are intended to provide a comprehensive means of evaluating whether any discharge of fill is environmentally acceptable. The standard for issuance of a 404 permit is consideration of the full public interest by balancing the favorable impacts of a proposed activity against the detrimental impacts to reflect the national concerns for both the protection and utilization of important resources. This balancing test of probable impacts involves many factors, including conservation, economics, flood hazards, land use, navigation, energy and mineral needs, and, in general, the needs and welfare of the people.[4] The Corps has considerable discretion, and the weight of each factor changes with each permit application, depending on the factor's importance and relevance to the particular proposal. However, a discharge is categorically prohibited if it would significantly degrade water quality. In addition, no discharge may be allowed if there is a less environmentally damaging practicable alternative. Where there is no other alternative, the discharge may be allowed if the applicant has taken all practicable steps to minimize the amount of material discharged and to compensate for unavoidable impacts through mitigation.

Section 404 permits consist of two basic types: Individual permits for a particular site and nationwide (general) permits for categories of discharges that are similar in nature and have no more than minimal adverse impacts, individually and cumulatively, on the waters of the United States. If the discharge may have more than minimal impacts, an individual permit is required. Nationwide permits cover approximately 74,000 activities annually (about 90% of total Corps permits) and involve less regulatory burden and time than authorization by individual permits. Disposal of excess overburden associated with mountaintop removal mining has generally been permitted under Nationwide Permit 21 (NWP 21), which authorizes discharges from surface coal mining activities that result in no more than minimal impacts (site-specifically and cumulatively) to the aquatic environment.[5] The use of NWP 21 in conjunction with mountaintop mining in the Appalachian region has been controversial and has been challenged in litigation. The Obama Administration recently added new environmental restrictions on the use of NWP21 generally, while permanently prohibiting its use to authorize discharges to construct valleys fills, such as occurs in the Appalachian region (these developments are discussed below).

The U.S. Fish and Wildlife Service (FWS) also has responsibilities relevant to mountaintop removal mining. FWS implements and enforces the Endangered Species Act (35 U.S.C. §1531) and the Fish and Wildlife Coordination Act (16 U.S.C. §661), and under both laws, agencies proposing projects affecting U.S. waters are required to consult with FWS to ensure that fish and wildlife conservation and impacts on threatened or endangered species are considered. Coordination with FWS is required for both SMCRA and CWA permits.

Criticism and Legal Challenges to Mountaintop Mining

Because of the increase in valley fill disposal of mountaintop mining overburden in areas of Appalachia, the practice has drawn public attention and criticism. Critics say that, as a result of valley fills, streams and the aquatic and wildlife habitat that they support are destroyed by tons of rocks and dirt. Flow regimes are altered, increasing the likelihood and severity of floods, and the water quality downstream from fills also is significantly degraded. In addition, mountaintop removal can crack the walls and foundations of nearby homes; cause dust, noise, and vibration from blasting; collapse drinking water wells; and destroy nearby streams for fishing, hiking, swimming, or aesthetic pleasure. It also has forced the relocation of whole communities.[6]

Environmental groups argue that the practice of authorizing valley fills under Section 404 is unlawful because mining overburden is waste material which pollutes and destroys waterways, and impacts are far more than minimal, which is the standard for coverage by a nationwide permit.

The mining industry argues that mountaintop mining is essential to conducting surface coal mining in Appalachia. The poor stability of the soil surrounding coal deposits in this region makes it impossible to mine the coal using underground mining techniques. Waste disposal in valley fills is a necessary part of that activity because of the steep topography of the region, and they assert that mountaintop mining would not be economic or feasible if producers were restricted from using valleys for the disposal of mining overburden. Requiring Section 402 permits for these activities would effectively prohibit a broad range of mining which has been allowed by long-standing practice, they say.

Critics have been using litigation to challenge the practice. In a number of cases discussed here, environmental groups have been successful at the federal district court level in challenging issuance of permits for mountaintop mining projects, but each has been later overturned on appeal. Nonetheless, the criticisms also have prompted some regulatory changes, also discussed below.

In 1998, a West Virginia citizen group sued the state of West Virginia and the Corps for failure to prevent or enforce against environmental violations caused by mountaintop removal practices. The principal claim under SMCRA was that the state was failing to enforce OSM's 1983 buffer zone rule, which protects intermittent and perennial streams from disturbance by coal mining activities.[7] In addition, the lawsuit asserted that the Corps had been granting permits that allowed disposal of waste in waters of the United States through permits under the nationwide permit program that have greater than minimal adverse effects, individually and cumulatively, which plaintiffs argued is contrary to the CWA. Some of the claims were settled when the federal agencies agreed to complete a Programmatic Environmental Impact Statement (PEIS) on the effects of mountaintop removal mining. The Corps also agreed that proposed valley fills in West Virginia in watersheds of at least 250 acres must be permitted by individual, not nationwide, permits.

The remaining claims were addressed in an October 1999 ruling which held that disposal of mining spoil in valley streams violates federal and state mining rules and the CWA.[8] Under the ruling, mining spoil was reclassified from "dredge and fill material," requiring a CWA Section 404 permit, to "waste material" that is subject to CWA Section 402 permit requirements, thus raising the regulatory hurdles for disposing of mining waste.

Upon appeal, the district court ruling was overturned in a decision that dealt with jurisdiction and state sovereignty issues.[9] The court held that the stream buffer regulation at issue was, in fact, a matter of state law, not federal law and, thus, the case should not have been brought in federal court. The Clinton Administration had sided with the industry by appealing the district court's finding that mountaintop mining must be regulated under CWA Section 402, but it concurred with the related finding, supported by environmental groups, that the activity violates stream buffer zone requirements under SMCRA. In 2002, the Supreme Court declined to hear a challenge to the Fourth Circuit decision.

In October 2005, the Corps, EPA, and other federal agencies released a final PEIS on the impacts of mountaintop mining and valley fills,[10] as promised in the 1999 partial settlement of *Bragg*. It identified three alternatives for improving coordination of regulatory efforts to limit the negative impacts of mountaintop mining. Under the preferred alternative, OSM, the Corps, EPA, and state agencies would determine the size, number, and location of valley fills for a proposed operation, under a joint permit application integrating the CWA and SMCRA programs. The Corps would make case-by-case determinations whether a project would be covered under NWP 21 or under an individual Section 404 permit. More than 70,000 public comments were submitted on the draft PEIS. Industry groups favored continued use of general permit authorizations, while environmental groups said that the preferred alternative failed to place adequate limits on mountaintop mining and valley filling. A number of comments were critical that all of the alternatives were process alternatives, and none would minimize the environmental impacts from valley fills. The agencies responded that the alternatives were appropriate for a programmatic EIS and that they would provide increased environmental protection. The agencies also said that a number of changes to agency rules, policy, and guidelines would follow.

The "Fill Material" Rule

Controversies also arose following a proposal by EPA and the Corps in 2000 to revise a portion of the regulations that implement CWA Section 404. The agencies proposed to redefine the terms "fill material" and "discharge of fill material." One result of the proposal would be regulatory definitions more consistent with the Administration's position in the then-ongoing *Bragg* litigation, namely its view that regulating mountaintop removal mining under

CWA Section 404 is not inconsistent with that act. This proposal was not finalized before the Clinton Administration left office but was finalized in May 2002 by the Bush Administration, substantially as proposed in 2000.[11]

The revision was intended to clarify the regulatory definition of fill material—which determines whether the activity is subject to Section 404 permit requirements or more stringent Section 402 requirements—by replacing two separate and inconsistent definitions with a single, common definition to conform with long-standing Corps and EPA practice in regulating surface mining activities. According to the Clinton and Bush Administrations, the previous definitional differences had led to considerable confusion, as reflected in part in the *Bragg* and other lawsuits, but the changes were not driven solely by concerns over regulating mountaintop mining practices. Environmental groups continue to contend that the disposal practice is unlawful under the Clean Water Act, and that the revised EPA and Corps rules allow for inadequate regulation of disposal activities, including coal mining waste.

Following issuance of the revised rule, the Senate Environment and Public Works Committee held an oversight hearing in the 107th Congress to examine the rule, receiving testimony from Administration, mining industry, and public witnesses. Legislation intended to reverse the 2002 rule has been introduced in each Congress since then (see "Congressional Actions" below).

OSM's 2008 Stream Buffer Zone Rule

The 2003 draft PEIS called for OSM to make changes to its stream buffer zone rule to improve consistency with the Clean Water Act, and OSM proposed changes to that rule in 2004. However, OSM subsequently decided to prepare a new PEIS and to draft a revised rule, both of which were released in 2007.[12]

OSM issued a final revised buffer zone rule in December 2008. As described by OSM, the final rule requires that surface coal mining operations be designed to minimize the amount of spoil placed outside the mined-out area, thus minimizing the amount of land disturbed. It also requires that, to the extent possible, surface coal mining and reclamation operations be designed to avoid disturbance of perennial or intermittent streams and the surface of lands within 100 feet of those streams. If avoidance is not reasonably possible, the rule requires that the permit applicant develop and analyze a range of reasonably possible alternatives and select the one that would have the least overall adverse impact on fish, wildlife, and related environmental values.[13] According to OSM, the final rule does not mandate avoiding placement of coal mine waste in or within 100 feet of perennial or intermittent streams in all cases, because "there is sometimes no viable alternative to the construction of coal mine waste disposal facilities in perennial or intermittent streams and their buffer zones, in which case avoidance is not reasonably possible."[14]

The 2008 revised rule eliminated the provision in the 1983 stream buffer zone rule that had required a finding that the proposed activity would not cause or contribute to a violation of state or federal water quality standards. In doing so, OSM said that the previous language more closely resembled the CWA than the underlying provisions of SMCRA. Because the SMCRA rule does not substitute for or supersede the CWA, mine operators still must comply with the requirements of that law.

Both industry and environmental groups said that the final rule did little to change the existing practice of disposing mountaintop mining spoil into valleys and streams. In fact, OSM stated that a key purpose of the rule was to reduce confusion about the 1983 rule and to conform the regulation to historic practice of federal and state authorities. Environmental groups said that the final rule would actually reduce environmental protection for streams by making it easier for coal mine operators to obtain exemptions from the stream buffer zone requirement, thus increasing destructive mining activities in and around streams; a coalition of these groups filed a lawsuit challenging the rule.[15] The Obama Administration requested that the federal court hearing this case vacate the 2008 stream buffer zone rule and remand it to the department, in order to return immediately to the more stringent 1983 rule until a replacement rule can be adopted. However, the court rejected the Administration's request in August 2009[16] (see discussion below, "The Stream Buffer Zone Rule").

Other Litigation

In other litigation challenging authorization of a specific mountaintop mining operation in Kentucky (rather than the general practice), a federal district court ruled in 2002 that the disposal of waste from mountaintop mining into U.S. waters is not allowed under Section 404, and the court permanently enjoined the Corps from issuing Section 404 permits for the disposal of mountaintop mining overburden where the purpose is solely to dispose of waste. In January 2003, a federal court of appeals ruled that the district court's action was too broad and lifted the injunction prohibiting the Corps from issuing Section 404 permits for disposal of mountaintop mining waste.[17]

In 2007, individual permits for four mountaintop mining operations in West Virginia were overturned by a federal district court. The court found that the probable impacts of the valley fills would be significant and adverse, that the mitigation plans for each permit were not sufficient to compensate for those adverse impacts, and that the Corps inadequately evaluated the cumulative impacts of the projects. The Corps appealed the court's orders, and in February 2009, the court of appeals reversed and vacated the district court's actions.[18] The court found that the Corps had not acted arbitrarily or capriciously in its evaluation of the projects' impacts, and it found the Corps' proposed mitigation plans sufficient for purposes of complying with the National Environmental Policy Act (NEPA). One judge on the panel wrote in dissent that in his view the Corps had failed to establish that the projects will have no significant adverse environmental impact, and thus the agency had not satisfied the requirements of NEPA. While this case was pending, uncertainty over its outcome had effectively blocked the Corps from issuing Section 404 permits in Appalachia. Thus, the ruling by the court of appeals allowed for resumption of permitting activities in the region.

Citizen groups also have filed lawsuits seeking generally to halt the Corps' use of Nationwide Permit 21 for mountaintop mining operations. In the first such case, a federal district court ruled that NWP 21 violates the Clean Water Act by authorizing activities that have more than minimal adverse environmental effects. The court enjoined the Corps from using NWP 21 to authorize new mountaintop mining in southern West Virginia and ordered the Corps to revoke previous authorization for 11 operations. On appeal, the judgment of the district court and the injunction against NWP 21 were vacated when the court of appeals found that the Corps had complied with the Clean Water Act when it promulgated NWP 21.[19]

In another recent case, a U.S. district court ruled in March 2009 that, when the Corps issued Nationwide Permit 21, its analysis of cumulative impacts was inadequate and its reliance on compensatory mitigation in determining the environmental impacts of valley fills was arbitrary and capricious.[20] The court again enjoined the Corps from using NWP 21 to authorize mountaintop mining activities in the Southern District of West Virginia. The decision requires that mining operations operating under Nationwide Permit 21 be halted in that region, but mining companies can seek individual permits from the Corps or appeal the decision. In June 2009, the Obama Administration filed a notice of appeal of the district court's ruling, which officials said was done for procedural reasons, as no policy decisions had been made. At the same time, the Administration announced a series of administrative actions regarding surface coal mining operations in Appalachia, including use of Nationwide Permit 21.

ADMINISTRATIVE ACTIONS AND CONGRESSIONAL ACTIVITY

The Obama Administration has joined the debate over mountaintop mining. Early in 2009, EPA began reviewing CWA Section 404 permit applications for surface coal mining operations in the Appalachian states, many of which had been on hold for months in light of the litigation on which the U.S. Court of Appeals for the Fourth Circuit ruled in February 2009, *Ohio Valley Environmental Coalition v. Aracoma Coal* (see above, "Other Litigation"). Following that ruling, EPA Administrator Lisa Jackson announced that, under its CWA authority to comment on 404 permit applications pending at the Corps and its authority to oversee issuance of Section 402 permits by states, EPA would review pending surface coal mining permit requests in Appalachia to ensure protection of the environment.

On June 11, 2009, officials of EPA, the Corps, and the Department of the Interior signed a Memorandum of Understanding (MOU) and Interagency Action Plan (IAP) outlining a series of administrative actions to reduce the harmful environmental impacts of mountaintop mining. The plan includes a series of near-term and longer-term actions that emphasize specific steps, improved coordination, and greater transparency of decisions.[21] Many of the Administration's actions have been highly controversial, resulting in substantial congressional criticism. Some have been challenged successfully in court by industry companies and groups, and other legal challenges continue.

The ECP

Also on June 11, 2009, EPA and the Corps signed a specific agreement detailing criteria to be used to coordinate and expedite review of pending permit applications for surface coal mining operations in Appalachia (including but not limited to mountaintop mining projects). Based on its authority to evaluate 404 permit applications, EPA identified 79 projects for additional environmental review. EPA and the Corps developed a joint Enhanced Coordination Procedure (ECP) for evaluation of the permits. Of the 79 projects, 49 were located in Kentucky, 23 in West Virginia, 6 in Ohio, and 1 in Tennessee. EPA stated that each of the 79 projects, as proposed, is likely to result in significant harm to water quality, either individually or cumulatively. Under the coordination process, the Corps was responsible for beginning discussions with EPA and the mining companies to reduce anticipated environmental effects, and generally the individual reviews were expected to be completed within 60 days of notification by the Corps. As of September 2011, 50 of the 79 permit applications had been withdrawn without prejudice by the applicants, and the Corps had issued permits for eight. EPA had sent recommendation letters for two others, and the 60-day review had begun for one more. The remaining 18 (all located in Kentucky and West Virginia) were pending, awaiting additional information from the applicant prior to Corps notification to begin the formal EPA review.

Coal industry groups and coal state officials contended that the ECP process resulted in costly delay in issuance of permits. They challenged the process in federal court, arguing that EPA and the Corps had violated the CWA by transferring regulatory authority from the Corps to EPA and effectively expanding the agency's veto power over permit decisions and also violated the Administrative Procedure Act by substantially changing permitting procedures without initiating a formal rulemaking. Further, they challenged the screening process used to determine whether pending permits should be subject to the ECP (called the Multi-Criteria Integrated Resource Assessment, or MCIR Assessment), as well as guidance used by EPA to evaluate the permit applications (discussed below, "EPA Guidance on Permitting"). EPA has defended the ECP, saying that the additional coordination between the two agencies has resulted in surface coal mining projects with reduced environmental, water quality, and human health effects, consistent with requirements of the CWA.

On October 6, 2011, the court struck down the ECP and the MCIR Assessment as an unlawful transfer of legal authority from the Corps to EPA.[22] Following the ruling, the Corps and EPA issued memoranda to clarify the roles of both agencies moving forward. While the specific ECP was set aside, the memoranda stated, the Corps will continue to process applications—including the 21 pending under the vacated ECP process—in accordance with permitting regulations, and EPA Regions should continue collaborating with the Corps to review proposed 404 permits.[23]

The Stream Buffer Zone Rule

OSM's 2008 stream buffer zone rule (discussed above) also was addressed in the June 2009 IAP. The Department of the Interior asked a federal court to vacate the revised rule that was issued by the Bush Administration in December 2008, saying that the rule does not adequately protect water quality and stream habitat. OSM officials had hoped to return to using the more stringent 1983 rule until, in the longer term, a replacement rule can be developed. As noted above, the court rejected this request, thus leaving the 2008 rule in place until a new regulation is issued.

However, OSM officials also have stated that because of litigation over the 2008 revisions, states were never directed to amend their programs to conform with the revisions—thus, states are still operating under the 1983 rule.

In November 2009, OSM identified a broad set of regulatory options that it is considering for revisions to the 2008 rule, ranging from formally reinstating the 1983 rule with small conforming changes, to requiring stricter buffer zone requirements for mountaintop mining operations on steep slopes.[24] OSM officials have been working on developing a new rule and an accompanying draft environmental impact statement (EIS). The revised rule is expected to apply nationwide, not just in Appalachia. OSM's efforts to revise the rule have been criticized, based on concerns about potential economic impacts of the rule and the quality of work on its EIS. However, OSM officials and environmental advocacy groups contend that a new rule is needed to protect waterways from surfacing mining operations. OSM expects to release a draft rule and draft EIS concurrently, but the agency does not have a timetable for doing so. EPA and OSM also have pledged to strengthen oversight of state CWA and SMCRA permitting, regulation, and enforcement activities.

Nationwide Permit 21

As noted previously, disposal of excess overburden associated with mountaintop removal mining has generally been authorized by the Corps under Nationwide Permit 21 (NWP 21), which authorizes surface coal mining discharges with no more than minimal impacts to the aquatic environment. As part of the IAP, in July 2009, the Corps published a two-part proposal concerning the use of NWP 21 to authorize mountaintop mining activities.[25] First, the Corps proposed permanent modification of this NWP to prohibit its use in conjunction with surface coal mining activities in the Appalachian region. Second, because modification of the NWP is a longterm process, the Corps also proposed to temporarily suspend NWP 21 for surface coal mining activities in the Appalachian region in order to quickly halt the use of NWP 21 in the region. Surface coal mining activities in other regions would not be affected. The proposed suspension and modification would mean that surface coal mining activities in Appalachia would need to be evaluated through the Corps' detailed individual permit review process, rather than under a streamlined nationwide permit. The Corps explained its reason in the proposal:

> [T]he Corps now believes that impacts of these activities on jurisdictional waters of the United States, particularly cumulative impacts, would be more appropriately evaluated through the individual permit process, which entails increased public and agency involvement, including an opportunity for public comment on individual projects.[26]

Subsequently, in June 2010, the Corps acted to suspend use of NWP 21 in the Appalachian region immediately (the second part of the July 2009 proposal) as an interim measure while continuing to evaluate permanent modification or suspension of the permit. According to the Corps' announcement, NWP 21 activities in the affected region that had been verified by the Corps prior to June 18, 2010, would continue to be authorized until March 18, 2012, unless modified on a case-by-case basis.[27] The Corps estimated that the immediate suspension would affect approximately six operations that at the time were seeking to use NWP 21; they would then have to submit applications for individual permits to authorize their activities.

In March 2012, the Corps reissued all of the existing nationwide permits, with modification of a number of them, including NWP 21.[28] The previous version of NWP 21, issued in 2007, did not have any acreage or linear foot limits and relied on permit conditions and pre-construction notification reviews to reduce adverse impacts on the aquatic environment. The Corps determined that this approach has not adequately protected against loss of aquatic resources; thus the reissued permit adds a 1/2-acre and 300-linear foot limit for the loss of stream beds when NWP 21 is used. Further, the reissued permit strictly prohibits use of NWP 21 to authorize discharges of dredged or fill material into U.S. waters to construct valley fills associated with surface coal mining. Projects no longer eligible under NWP 21 may seek authorization under a Section 404 individual permit, which can be issued for longer periods of time than a nationwide permit. The effective date of the reissued NWPs is March 19, 2012; they will expire on March 18, 2017.

EPA Guidance on Permitting

Throughout the Administration's review of activities that began in 2009, mining industry representatives have vigorously criticized EPA's actions and questioned what criteria are being used for permit reviews, saying that the agencies are needlessly delaying important projects, thus negatively affecting both jobs and the nation's energy security. The mining industry, environmental advocates, and elected officials in the affected region all had been asking for greater clarity. EPA and the Corps committed to issue guidance to strengthen environmental review of surface coal mining permits and operations and to clarify applicable criteria.

On April 1, 2010, EPA released a 31-page interim guidance memorandum to clarify the agency's tightened requirements for surface coal mining in Appalachia. The guidance would be applied as a framework for EPA's approval of all pending and future reviews for permits to dispose of coal mining waste—through CWA Section 404 permits—and other types of discharges from Appalachian surface coal mining that are authorized by Section 402/NPDES permits. According to EPA, the guidance is not intended to bring a complete halt to mountaintop removal mining, but it should force the industry to adopt a practice of minimal or zero filling of valleys with mining debris. In addition to applying the guidance to reviewing 404 permits, since late 2010, EPA has cited the scientific evidence in the guidance in comments and some objections to proposed NPDES permits for coal mining projects in Kentucky and West Virginia.

Among other items, the document set strict numeric limits on conductivity levels in waters affected by mining activities and valley fills. Conductivity is a measure of the level of salinity in water, which is expressed as microSiemens per centimenter, or μS/cm. As conductivity levels rise, fish, amphibians, mussels, and other aquatic organisms can be adversely affected. In the guidance (and in its ongoing review of pending Section 404 and 402 permit applications), EPA refers to increasing scientific recognition of a strong relationship between elevated total dissolved solids and conductivity levels in Appalachian streams and adverse impacts to aquatic life in streams and rivers below surface coal mining operations. Based on its review of recent scientific literature, EPA has concluded that, as a general matter, where conductivity levels will exceed 500 μS/cm in central Appalachia, there is a reasonable potential that aquatic biota will be adversely affected and applicable water quality standards will be violated, thus establishing this as a likely upper limit to be allowed in permits. Environmental groups support EPA's use of conductivity to assess water quality impacts of coal mining, but industry groups have been highly critical, asserting that the science linking conductivity to water quality impairment is uncertain and that acceptable numeric levels are arbitrary. While the 2010 interim guidance was effective immediately, EPA also sought public comment to inform possible future changes to the guidance. Conductivity, and its use in assessing coal mining impacts on water quality, has become a focus of debate and public comment.

The EPA interim guidance also referenced two draft reports produced by EPA's Office of Research and Development that were used, together with peer-reviewed science, in preparing the guidance. The first draft report assessed the state of the science on the environmental impacts of mountaintop mines and valley fills on streams in Central Appalachia. The second developed a conductivity benchmark value that is intended to protect the biological integrity of aquatic life in waters in the region.

At EPA's request, the agency's Science Advisory Board (SAB) conducted evaluations of the EPA scientific documents, in view of the reports' likely importance to regulatory actions. The panel agreed with EPA's overall conclusions that there is strong evidence for a causal relationship between mountaintop mining/valley fills and harm to downstream water quality and to biological impacts on organisms that live in and on the bottom of streams. The panel also concluded that there is clear evidence that valley fills are associated with increased levels of dissolved ions (measured as conductivity) in downstream waters and that these increased levels of conductivity are associated with changes in the composition of stream biological communities.

For both reports, the SAB panel made a number of recommendations to strengthen both the supportive and the cautionary aspects of its comments on the EPA draft documents. For example, the conductivity report derives a benchmark that is intended to avoid the extinction of native species exposed to mountaintop mining/valley fills in Appalachian streams. The SAB panel stated that because the complete loss of a genus is an extreme ecological effect, it may be appropriate to develop a more appropriate endpoint recognizing less severe, but still significant, effects that occur at lower exposure levels. A consequence of doing so could be EPA's adoption of a more protective conductivity benchmark than contained in the April 2010 guidance described above and application beyond Appalachian mountaintop mines, at least in areas where sufficient data exist. At the same time, the SAB told EPA that the reports need to be clarified to emphasize that both conductivity and total dissolved solids (both referring to an increase in salinity in otherwise dilute freshwater) are "coarse" indicators of water quality, because of confounding factors such as the presence of other elements, and should not be used exclusively as surrogates for water quality. Thus, the SAB supported conductivity along with other field-based tests and cautioned that EPA should not over-rely on conductivity tests; the agency needs to weigh a variety of environmental factors in its assessments. Based on the SAB reviews, EPA released final versions of the two scientific reports in March 2011.[29]

The April 2010 interim guidance has been very controversial. Environmental advocates defend EPA's implementation of the guidance as a way to protect streams from the harmful effects of surface coal mining in Appalachia. But industry and some states oppose EPA's efforts. Lawsuits were brought by the states of Kentucky and West Virginia, and were consolidated with the litigation brought by coal companies and trade associations over the ECP process (*National Mining Association v. Jackson*, discussed above). They argue that the conductivity benchmark is set below background levels for some healthy streams in the region, making it too difficult to achieve. The lawsuits also argue that the guidance is an attempt to write new rules without following the notice-and-comment requirements of the Administrative Procedure Act. In January 2011, a federal judge denied industry's request to block implementation of the guidance, but also denied the government's request to dismiss the case.[30]

EPA released final guidance to improve the agency's review of Appalachian surface coal mining operations on July 21, 2011.[31] The final document retains the same numeric limits on conductivity as in the interim guidance. However, EPA officials describe the final guidance as including considerable flexibility, based on considering public comments and experience in implementing the interim guidance since April 2010. For example, the final guidance reflects the SAB's recommendations that the specific conductivity limits not be used outside waters of West Virginia and Kentucky, until the limits can be scientifically verified as applicable to other waters. Further, the final guidance acknowledges that at sites where the strict conductivity limits cannot be attained, EPA may recommend alternative permit conditions such as limits on individual ions, specifying particular best management practices for mining project design to minimize harmful impacts, or allowing for offsets in some circumstances. Environmental groups endorsed the final guidance, saying that it is key to strengthening permitting processes. Industry groups continue to argue that EPA has unlawfully expanded its statutory authority and infringed on the role of states.

On July 31, a federal judge issued summary judgment for mining industry and state plaintiffs who had challenged EPA's use of the guidance.[32] The court rejected the government's arguments that the guidance is a nonbinding document to help EPA assess whether a mine's actions would violate CWA standards. Instead, the court agreed with plaintiffs' contentions that the guidance is being implemented as binding and having a practical effect on the permitting process for new Appalachian surface coal mining projects. Further, the court concluded that the guidance impermissibly sets a conductivity criterion for water quality standards, overstepping authority that the CWA assigns to states. The government is expected to appeal both this ruling and the same court's October 2011 invalidation of the ECP process.

EPA Veto of a Mountaintop Mining Permit

In addition to directing EPA to issue the environmental guidelines used by the Corps to evaluate permit applications (CWA Section 404(b)), Section 404 also authorizes EPA to prohibit or otherwise restrict the specification by the Corps of a site for the discharge of dredged or fill material, if the agency determines that the activity will have an unacceptable adverse effect on water supplies, fish, wildlife, or recreational areas. EPA has used this veto authority, under Section 404(c), only 12 times since 1972. None of the previous vetoes involved a surface coal mining or mountaintop mining project.

On January 13, 2011, the agency reached a determination to veto a permit for a mountaintop mining operation in West Virginia. According to EPA, the Spruce No. 1 mine in Logan County, West Virginia, as proposed, would be one of the largest surface mining operations ever authorized in Appalachia, and waste disposal from the mine would bury over seven miles of streams, directly impact 2,278 acres of forestland, and degrade water quality in streams adjacent to the mine. The Corps issued a permit for the project in 2007, but it was subsequently delayed by litigation and has been operating on a limited scale since then. EPA acknowledges that the project has been modified in order to reduce impacts, but the veto determination is based on the agency's conclusion that the project could result in unacceptable adverse impacts to wildlife and fishery resources.[33] Some of the water quality problems cited in EPA's veto of the permit are from increased loads of pollutants that the project would discharge—particularly selenium and total dissolved solids—which contribute to increased stream conductivity, the same adverse impact cited by EPA in the permitting guidance document discussed above.

EPA's veto of the permit has been very controversial, in part because it involves cancelling a permit previously issued by the Corps.[34] Coal industry groups and organizations representing manufacturing and other sectors have been highly critical of EPA's actions, many saying that to revoke an existing permit creates huge uncertainty about whether water quality permits would be rescinded in the future, producing a ripple effect beyond the coal industry.

The owner of the site, the Mingo Logan Coal Company, challenged EPA's action in federal court, even before the veto was finalized. EPA argued that the veto, while highly unusual, is justified because the project involves unacceptable environmental impacts.

The agency said that it is not currently reviewing any other previously authorized Appalachian surface coal mining project pursuant to Section 404(c). In March, a federal district court agreed with the industry petitioners and concluded that the CWA does not give EPA the power to render a permit invalid once it has been issued by the Corps. Although the language of 404(c) is "awkwardly written and extremely unclear," the court found EPA's view that it has such authority an unreasonable interpretation of the statute.[35] Thus, it overturned the veto.

The government has appealed the district court's ruling, and a streamlined briefing schedule has been set by the U.S. Court of Appeals for the D.C. Circuit.

Viewed broadly, the Administration's combined actions on mountaintop mining displease both industry and environmental advocates. The additional scrutiny of permits, more stringent requirements, and EPA's veto of a previously authorized project have angered the coal industry and many of its supporters. At the same time, while environmental groups support the veto and EPA's steps to restrict the practice, many favor tougher requirements or even total rejection of mountaintop mining in Appalachia.

So far, the Administration has not fared well when federal courts have reviewed its stepped-up actions involving Appalachian surface mining. As described above, courts have ruled that OSM exceeded its authority in seeking to revise the stream buffer rule without complete administrative procedures, and that EPA exceeded its authority by using the Enhanced Coordination Process (the ECP) and the agency's 2011 permitting guidance, as well as by its veto of the Spruce No. 1 mine permit.

Congressional Actions

Congressional interest in these issues also has been evident and has increased substantially in the 112th Congress.

In the 112th Congress (as in several prior Congresses), legislation to reverse the 2002 revised regulations that define "fill material," discussed previously, has been introduced (H.R. 1375, the Clean Water Protection Act). This bill would sharply restricted mountaintop mining by excluding from the definition of "fill material" any pollutant that is discharged into water primarily for the purpose of disposing of waste. This provision would allow pollutant discharges that replace portions of the waters of the United States with dry land or which change the bottom elevation of a water body for any purpose to be considered fill material. But it would reject the view reflected in the 2002 regulations that some discharges for purposes of waste disposal (including mine overburden) should be allowable within the definition of fill.

Legislation similar to H.R. 1375 was introduced in the 111th Congress (H.R. 1310). A somewhat narrower legislative approach was contained in another bill in the 111th Congress, the Appalachia Restoration Act (S. 696). It was similar to H.R. 1310 in that it would have defined fill material to include pollutant discharges that replace portions of the waters of the United States with dry land or which change the bottom elevation of a water body for any purpose. But it would have excluded the disposal of excess spoil material from coal surface mining and reclamation activities, as described in Section 515(b)(22) of SMCRA, in waters of the United States. This provision appeared to allow discharges from some mining practices to be considered fill material, such as hardrock mining or mining of other minerals such as sand and gravel (thus qualifying for a Section 404 permit), while excluding discharges from surface coal mining activities from the definition of fill material.

In June 2009, the Senate Environment and Public Works Subcommittee on Water and Wildlife held a hearing on the impacts of mountaintop removal coal mining on water quality in Appalachia. Public witnesses, including an environmental scientist from the University of Maryland's Center for Environmental Science, described the impacts of mountaintop removal with valley fills as immense and irreversible. An EPA witness said that the agency must reinvigorate its oversight of state-delegated CWA programs, such as the Section 402 permit program, to address water quality issues before Section 404 decisions are made. A witness from the West Virginia Department of Environmental Protection said that many in that state are concerned that ongoing administrative and possible legislative actions could limit economic development opportunities for all types of mining, not just mountaintop mining.

The significance of both bills in the 111th Congress was that discharges of materials that are not eligible for a Section 404 permit are regulated under CWA Section 402. Because Section 402 discharge requirements are more restrictive than those for Section 404, some discharges that could be permitted under Section 404 cannot be authorized under Section 402. Supporters of the bills favored making it more difficult to use Section 404 to authorize activities that they consider to be environmentally harmful. On the other hand, critics said that, as a practical matter, economically important activities such as coal mining could not meet the more stringent limitations of a Section 402 permit and, thus, would be infeasible.

Additionally, legislation intended as criticism of the Obama Administration's regulatory actions on Appalachian coal mining also was introduced in the 111th Congress (H.R. 6113, and S. 3933 , the Electricity Reliability Protection Act of 2010). This bill would have prohibited EPA, the Army Corps, and OSM from administering or enforcing any policy or procedure that was announced in the June 2009 MOU or the April 2010 EPA permitting guidance unless they are contained in promulgated regulations. Critics of the Administration's actions have argued that the policies constitute rules, and thus should be subject to complete administrative requirements of rulemaking, including public notice and comment and subsequent judicial review.

In the 112th Congress, similar legislation has not been introduced, but was included as an amendment to a House-passed bill providing FY2011 appropriations for EPA and other agencies (H.R. 1). The amendment would have prohibited funding for a number of EPA regulatory activities discussed previously in this report, including implementation of the regulatory initiatives on mountaintop mining in Appalachia.[36] However, the full-year FY2011 appropriations bill enacted in April 2011 did not include that provision (P.L. 112-10). Several similar provisions were included in FY2012 appropriations bills considered by the House (H.R. 2354, providing funding for the Corps, and H.R. 2584 to fund EPA), but none of the language was included in an omnibus appropriations bill providing full-year FY2012 funding for EPA, the Corps, and other agencies that Congress enacted in December 2011 (P.L. 112-74).

As noted above, EPA's January 2011 veto of a Section 404 permit for the Spruce No. 1 mine in West Virginia—overturned by a federal court in March 2012—has been very controversial, including in Congress. In the 111th

Congress, legislation was introduced to delete Section 404(c) from the CWA (H.R. 5992), thus removing EPA's authority to veto permits for projects, but no further action occurred. Similar legislation has been introduced in the 112th Congress (H.R. 517). In addition, other bills have been introduced that are intended to address the veto of the Spruce No. 1 mine project. One proposal would bar EPA from using the 404(c) authority "after the fact," that is, after the Corps has issued a 404 permit (H.R. 457/S. 272). A second proposal would similarly prohibit retroactive vetoes and also would set deadlines for EPA's 404(c) authority and clarify procedures for elevating 404 permitting decisions to higher level agency and department officials, based on consultation with and consent of affected states (S. 468/H.R. 960). A third proposal (H.R. 2018) would modify a number of CWA provisions; one would prohibit EPA from vetoing a 404 permit without concurrence of the state in which the discharge originates and would require EPA to submit any comments on a Corps-proposed 404 permit within 60 days (currently there is no statutory time limit); the House passed this bill on July 13, 2011.

The House Transportation Subcommittee on Water Resources and Environment held oversight hearings on these issues in May 2011, and the House Oversight and Government Reform Subcommittee on Regulatory Affairs held similar hearings in July 2011 and July 2012.

OSM's efforts to revise the 2008 stream buffer zone rule also have been controversial. The House Natural Resources Committee has held several hearings to air concerns that the expected rewrite of the rule would have harmful job impacts on surface coal mining. In February, the Committee approved legislation (H.R. 3409) to limit the authority of the Secretary of the Interior to issue a revised rule. Similar legislation has been introduced in the Senate (S. 3450) and also is included as a provision of another Senate bill (S. 3445).

The Obama Administration did not present formal views on any of the legislation in the 111th Congress, but the Administration did express strong opposition to H.R. 2018 in the 112th Congress.[37]

End Notes

[1] Nancy Stoner, Acting Assistant Administrator for Water, and Cynthia Giles, Assistant Administrator for Enforcement and Compliance Assurance, Improving EPA Review of Appalachian Surface Coal Mining Operations under the Clean Water Act, National Environmental Policy Act, and the Environmental Justice Executive Order, U.S.

Environmental Protection Agency, Memorandum, July 21, 2011, p. 5, http://water.epa.gov/ lawsregs/guidance/wetlands/upload/ Final_Appalachian_Mining_Guidance_072111.pdf.

[2] In the Appalachian states where mountaintop mining occurs, the SMCRA regulatory program has been delegated by the federal government to state agencies, except in Tennessee.

[3] The CWA authorizes delegation of both of these permit programs to qualified states. The NPDES program has been delegated to 46 states, including each of the Appalachian states. The Section 404 program has been delegated to two states, Michigan and New Jersey.

[4] 33 CFR § 320.4(a)(1).

[5] For additional information, see CRS Report 97-223, The Army Corps of Engineers' Nationwide Permits Program: Issues and Regulatory Developments, by Claudia Copeland.

[6] Daniel L. Rosenberg, "Mountaintop Mining and Proposed Rule Change Will Waste Clean Water Act," National Wetlands Newsletter, vol. 22, no. 4, July-August 2000, p. 12.

[7] In order to protect streams from sedimentation and channel disturbance, the 1983 buffer zone rule provided that no land within 100 feet of a perennial or intermittent stream shall be disturbed by surface mining activities, including the dumping of mining waste, unless the regulatory authority grants a variance that specifically authorizes surface mining activities closer to or through such a stream. The regulatory authority must find that the proposed mining activity will not cause or contribute to a violation of applicable state or federal water quality standards and will not adversely affect water quantity and quality or other environmental resources of the stream.

[8] Bragg v. Robertson, 72 F.Supp.2d 642 (S.D.W.Va. 1999).

[9] Bragg v. Robertson, 248 F.3d 275 (4th Cir. 2001).

[10] http://www.epa.gov/Region3/mtntop/eis2005.htm. The final PEIS was preceded by a draft PEIS issued in 2003.

[11] Department of the Army, Corps of Engineers, and Environmental Protection Agency, "Final Revisions to the Clean Water Act Regulatory Definitions of 'Fill Material' and 'Discharge of Fill Material,'" 67 Federal Register, No. 90, May 9, 2002, pp. 31129-31143. For additional background information on the 2002 rule, see CRS Report RL31411, Controversies over Redefining "Fill Material" Under the Clean Water Act, by Claudia Copeland.

[12] 72 Federal Register 48890, August 24, 2007.

[13] Department of the Interior, Office of Surface Mining Reclamation and Enforcement, "Excess Spoil, Coal Mine Waste, and Buffers for Perennial and Intermittent Streams; Final Rule," 73 Federal Register 75814-75885, December 12, 2008, p. 75875.

[14] Ibid., p. 75833.

[15] Coal River Mountain Watch v. USA, case no. 1:08-cv-02212-HHK C, D.D.C., filed December 22, 2008.

[16] Nat'l Parks Conservation Ass'n v. Salazar, 660 F. Supp. 2d 3 (D.D.C. 2009).

[17] Kentuckians for the Commonwealth v. Rivenburgh, 317 F.3d 425 (4th Cir. 2003).

[18] Ohio Valley Environmental Coalition v. Aracoma Coal Company, 556 F.3d 177 (4th Cir. 2009).

[19] Ohio Valley Environmental Coalition v. Bulen, 429 F.3d 493 (4th Cir. 2005).

[20] Ohio Valley Environmental Coalition v. Dana R. Hurst, 604 F. Supp. 2d 860, (S.D. W.Va. 2009).

[21] For additional information on these Administration actions, see http://water.epa.gov/lawsregs/ guidance/wetlands/ mining.cfm.

[22] National Mining Association v. Jackson, 816 F.Supp. 2d 37 (D.D.C. 2011). This ruling only addressed the ECP and the MCIR Assessment decision tool. A second part of the litigation, addressing the EPA guidance document, is pending.

[23] Nancy Stoner, Acting Assistant Administrator, U.S. EPA Office of Water, "Transmittal of Department of the Army Memo on 'Decision in National Mining Association et al. v. Jackson, et al,'" http://water.epa.gov/lawsregs/guidance/ wetlands/upload/EPA_ECP_ memo_DCCruling_11-16-11.pdf.

[24] Department of the Interior, Office of Surface Mining Reclamation and Enforcement, "Stream Buffer Zone and Related Rules," 74 Federal Register 62664-62668, November 30, 2009.

[25] Department of Defense, Department of the Army, Corps of Engineers, "Proposed Suspension and Modification of Nationwide Permit 21," 74 Federal Register 34311-34316, July 15, 2009.

[26] Id. at 34313.

[27] Department of Defense, Department of the Army, Corps of Engineers, "Suspension of Nationwide Permit 21," 75 Federal Register 117, June 18, 2010.

[28] Department of Defense, Department of the Army, Corps of Engineers, "Reissuance of Nationwide Permits," 77 Federal Register 10184-10290, February 21, 2012. Under the CWA, the previous nationwide permits, which were issued in March 2007, expire after five years and must be renewed thereafter.

[29] The final EPA reports are entitled The Effects of Mountaintop Mines and Valley Fills on Aquatic Ecosystems of the Central Appalachian Coalfields (EPA/600/R-09/138F) and A Field-Based Aquatic Life Benchmark for Conductivity in Central Appalachian Streams (EPA/600/R-10/023F). Both are available at http://www.epa.gov/ncea. The SAB's reviews are available under "reports" at http://yosemite.epa.gov/sab/sabpeople.nsf/ Web Committees/BOARD.

[30] National Mining Association v. Jackson, 768 F.Supp. 2d 34 (D.D.C. 2011).

[31] See http://www.epa.gov/owow/wetlands/guidance/mining.html.

[32] National Mining Association v. Jackson, D.D.C., No. 1:10-cv-1220, July 31, 2012.

[33] U.S. Environmental Protection Agency, "Final Determination of the Assistant Administrator for Water Pursuant to Section 404(c) of the Clean Water Act Concerning the Spruce No. 1 Mine, Logan County, WV; Notice," 76 Federal Register 3126-3128, January 19, 2011. The Final Determination and related materials are available at http://water.epa.gov/lawsregs/ guidance/cwa/dredgdis/spruce.cfm.

[34] There is some ambiguity about whether it is the first EPA veto of a previously issued permit. Initially, EPA said that the agency had never done so, but agency officials subsequently have cited the 1981 veto of a project in North Miami Beach, FL, and a 1992 veto for a municipal water supply reservoir project. The facts of those cases differ from the Spruce No. 1 mine case and did not involve a surface mining project.

[35] Mingo Logan Coal Co. v. EPA, D.D.C., No. 1:10-cv-541, March 23, 2012.

[36] For information, see CRS Report R41698, H.R. 1 Full-Year FY2011 Continuing Resolution: Overview of Environmental Protection Agency (EPA) Provisions, by Robert Esworthy.

[37] Executive Office of The President, Office of Management and Budget, "Statement of Administration Policy," July 12, 2011, http://www.whitehouse.gov/sites/default/files/ omb/legislative/sap/112/saphr2018r_20110712.pdf.

In: U.S. and World Coal Production
Editor: Raul M. Schmidt
ISBN: 978-1-62618-976-8

Chapter 3

PROSPECTS FOR COAL IN ELECTRIC POWER AND INDUSTRY*

Richard J. Campbell, Peter Folger and Phillip Brown

SUMMARY

For most of the twentieth century, the primary use of coal in the United States was for electric power generation, and for most of the history of power generation in the United States, coal has been the dominant fuel used to produce electricity. Even as recently as 2011, coal was the fuel used for almost 42% of power generation in the United States accounting for 93% of coal use. Industrial uses represented the remaining 7%. However, in April 2012, coal's share of the power generation market dropped to about 32% (according to Energy Information Administration statistics), equal to that of natural gas. Coal was the fuel of choice because of its availability and the relatively low cost of producing electricity in large, coal-burning power plants which took advantage of coal's low-priced, high energy content to employ economies of scale in steam-electric production. However, coal use for power generation seems to be on the decline, and the magnitude of coal's role for power generation is in question. Two major reasons are generally seen as being responsible: the expectation of a dramatic rise in natural gas supplies, and the

* This is an edited, reformatted and augmented version of the Congressional Research Service Publication, CRS Report for Congress R42950, dated February 4, 2013.

impact of environmental regulations on an aging base of coal-fired power plants.

A recent drop in natural gas prices has been enabled by increasing supplies of natural gas largely due to horizontal drilling and hydraulic fracturing (i.e., fracking) of shale gas formations. If the production can be sustained in an environmentally acceptable manner, then a long-term, relatively inexpensive supply of natural gas could result. Decreased natural gas prices are lowering wholesale electricity prices, stimulating a major switch from coal to gas-burning facilities. The electric utility industry values diversity in fuel choice options since reliance on one fuel or technology can leave electricity producers vulnerable to price and supply volatility. However, an "inverse relationship" may be developing for coal vs. natural gas as a power generation choice based on market economics alone, and policies which allow one fuel source to dominate may come at the detriment of the other.

Coal-fired power plants are among the largest sources of air pollution in the United States. More than half a dozen separate Clean Air Act programs could possibly be used to control emissions, which makes compliance strategy potentially complicated for utilities and difficult for regulators. Because the cost of the most stringent available controls, for the entire industry, could range into the tens of billions of dollars, some power companies have fought hard and rather successfully to limit or delay regulations affecting them, particularly with respect to plants constructed before the Clean Air Act Amendments of 1970 were passed. The expected retirement of approximately 27 GW of coal-fired capacity by 2016 has been reported to the Energy Information Administration (EIA) by coal plant owners and operators, accounting for approximately 8.5% of U.S. coal-fired capacity. While the costs of compliance with new Environmental Protection Agency regulations are a factor, several other issues are cited by coal plant owners and operators as contributing to these retirement decisions including the age of coal-fired power plants, flat to modest electricity demand growth, the availability of previously underutilized natural gas combined-cycle power plants, and the lower price of natural gas due to shale gas development. Even coal plants which have made significant modifications to meet existing EPA regulations are being closed or mothballed due to a combination of low natural gas prices, and the inability to sell power into other markets.

EIA expects coal to be a significant part of the U.S. power generation industry's future to well past 2030. But given price competition from natural gas, and emerging environmental regulations, that role will likely be smaller than in recent decades. Coal-fired generation is likely to face a challenging future.

INTRODUCTION

Coal has been the fuel most used for thermal energy since the early days of the Industrial Revolution. However, the mechanization of factories quickly brought a shift in coal use from heat to steam engines, and then to fuel steam turbines to power the electric motors of industry. The advent of electrical lighting and the electrification of buildings and residences drove even greater demand for electricity. For most of the twentieth century, the primary use of coal in the United States was for electric power generation.

In 2011, 93% of coal consumed in the United States was for electric power production, according to the Energy Information Administration (EIA). **Figure 1** shows the use of the four major types of coal produced in the United States, with bituminous and subbituminous coal dominating electric power generation.

The various types of coal differ mostly in carbon content and heating value:

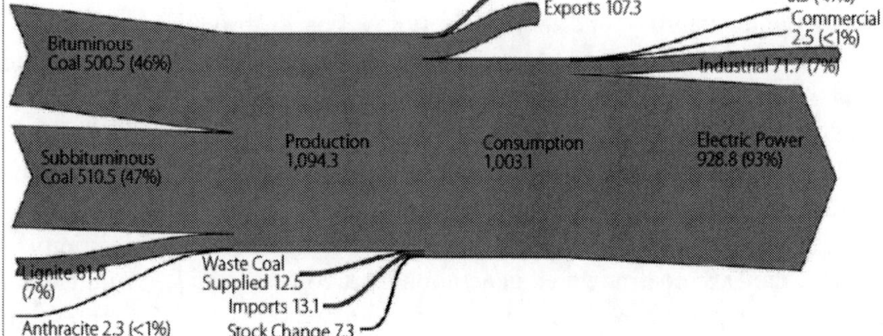

Source: U.S. Energy Information Administration. See http://www.eia.gov/totalenergy/data/annual/pdf/sec7_3.pdf.

Figure 1. Flow of U.S. Coal Consumption for 2011.

- **Anthracite** has the highest carbon content (between 86% and 98%), and a heat value of about 15,000 British Thermal Units (BTUs) per pound.[1] Anthracite coal is a small part of the electric power market, and is mostly found in the Appalachian region of Pennsylvania.
- **Bituminous** is the most abundant form of coal in the United States, and is the type most commonly used to generate electricity. Bituminous coal has a carbon content of approximately 45% to 86%, and a heat value between 10,500 BTUs and 15,500 BTUs.
- **Subbituminous** coal is mostly found in six western states and Alaska. It has a carbon content of between 35% and 45%, and a heat value of between 8,300 BTUs and 13,000 BTUs. Subbituminous coal generally has a lower sulfur content than other types of coal.
- **Lignite** has the lowest carbon content of the four types of coal generally used for electric power generation, averaging between 25% and 35%, and a high moisture and ash content. It also has the lowest heat value ranging between 4,000 BTUs and 8,300 BTUs.

Electric Power Generation

For most of the history of power generation in the United States, coal has been the dominant fuel used to produce electricity. Even as recently as 2011, coal was used to fuel almost 42% of power generation in the United States, as shown in Figure 2. Coal has been the fuel of choice for many decades because of its wide availability, and the relatively low cost of producing electricity in large, coal-burning power plants. Coal's low-priced, high energy content enabled the building of power plants able to take advantage of economies of scale in steam-electric production.

In a steam power plant, coal (or other fossil fuel) is burned to provide heat for turning water into steam in a boiler.[2] The steam is then forced under pressure into a steam turbine-driven generator which produces electricity. As of 2011, the U.S. coal-powered generation fleet consisted of 1,387 units with a nameplate capacity of almost 318 gigawatts[3] (GW)[4] representing approximately 30% of total U.S. generating capacity.[5] Electric utilities and independent power producers[6] are the primary entities involved in U.S. electricity production. Together, these electric power sector participants have combined for an average annual consumption of almost 995 million tons of coal in the period from 2002 to 2011,[7] with electric utilities consuming between 60% to 70% of the total (primarily in base load[8] power plants).

Net U.S. Generation 2011

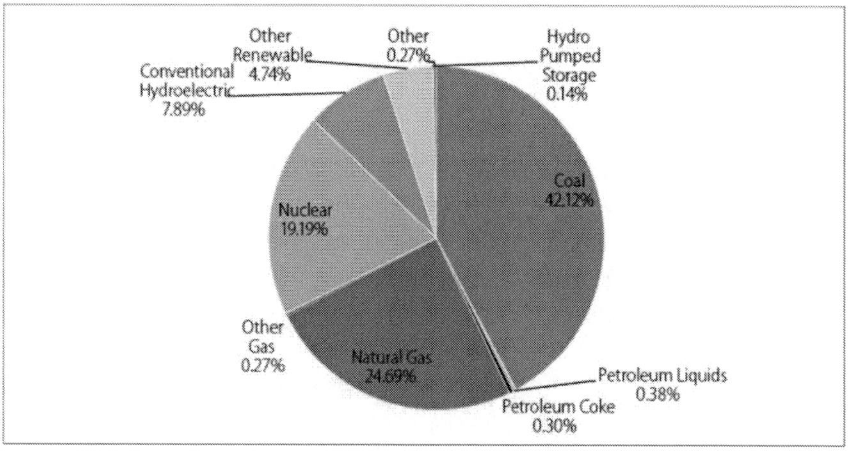

Source: CRS. Data from Energy Information Administration. Electric Power Monthly. Table 1.1, All sources of power.

Figure 2. Sources of Electric Power Generation.

The capital costs of power plant construction (i.e., the costs of finance and construction) vary according to the type (for example, coal or nuclear), and size of power plant. The cost for these plants to generate electricity is largely based on their cost of fuel, as well as operations and maintenance costs, other capital costs, and how the power plant is operated (for example, whether the facility is run as a base load or peaking[9] facility).[10] While the regulatory regime (i.e., under a rate-regulated or competitive environment), costs of transmission and distribution, and weather are strong determinants, the cost of fuel is often the major factor affecting electricity production especially for power plants which have already recovered their capital costs of construction.

Transportation costs for coal have historically been a major consideration in the choice of coal as a fuel. Barge transport is used where river access is available. But rail transportation is the most common mode of transit for coal as unit trains[11] offer the ability to transport large shipments to power plants on a regular basis. However, rail transportation can be costly.

> First, coal-burning electric power plants receive approximately 72 percent of their coal by rail. Second, rail transportation costs account for a sizable share of total delivered costs. While, on average, transportation costs account for approximately 20 percent of total [delivered] costs, they can reach as high as 59 percent on shipments of coal originating in the Powder River Basin.[12]

U.S. coal transportation costs are cited as a major reason why many utilities in the New England states have chosen to generate electricity with natural gas, or import coal from overseas as a lower cost alternative.[13]

Recent Changes in Markets for Electricity Fuels

The amounts of coal consumed for electricity generation from 2002 to 2011 are shown in **Figure 3**. While coal was the dominant fuel for electricity production in the period, the trend of increasing consumption seems to be slowing with a recent downturn in coal consumption for electricity beginning in about 2008. At the same time, natural gas use for electric power production is rising.

Increasing supplies of natural gas are largely behind the switch in fuel for power generation. Natural gas prices have been volatile over the last few decades, ranging from less than $2 per million Btus (MMBTU) to over $18 per MMBTU.[14] However, supplies of natural gas increased markedly beginning in 2008-2009 largely due to development of unconventional shale gas through improved drilling and simulation techniques. Horizontal drilling and hydraulic fracturing (i.e., fracking) have dramatically improved shale gas production, and the increased supplies of natural gas have caused prices to fall.

Power Plant Dispatch and Competitive Power Production

Many regions of the United States use competitive markets for power generation which sell into the wholesale market. These markets are generally administered by regional transmission organizations (RTOs) which have a goal of minimizing the cost of wholesale electricity to distribution utilities. The distribution utilities sell power to end-use (retail) consumers. In RTO markets, power plants bid against each other to serve the electricity demand of the wholesale market. Operators "dispatch" (or schedule the operation of) power plants based on the least costly choice, considering forecast loads and transmission system constraints for specific load increments at various times of the day or year. The most efficient power plants with the lowest heat rates (the efficiency with which a power plant turns fuel into electricity) are typically dispatched more, and lower efficiency, higher heat rate units are dispatched less.

Lower natural gas prices have caused wholesale electricity prices to decline, lowering potential revenues. However, many of the newer, more efficient natural gas combined cycle power generation plants have heat rates around 7,000 BTU/kWh (compared to today's coal plants with an overall average heat rate of 10,000 BTU/kWh). As natural gas prices have declined, natural gas plants have been dispatched more, and gross margins have increased with the spark spread (a measure of the potential profitability of a natural gas-fired power plant based on the difference between the price that a generator can obtain from selling one megawatt hour (MWh) of electricity and the cost of the natural gas needed to generate the MWh of electricity).

For coal plants, the lower cost of wholesale power has combined with the rising costs of coal to decrease their potential profitability. Older, less efficient coal power plants are especially at risk because wholesale prices of electricity may be inadequate for such facilities to operate economically.

Figure 4 shows natural gas and coal fuel costs for electricity production from 2002 to 2012. The right-most part of the chart is based on monthly data, illustrating the recent drop in natural gas prices for electricity production while coal costs continue to inch higher. The drop in natural gas prices is a major factor in the recent switch from coal to gas-burning facilities, since the cost of fuel is a major determinant in the dispatch of electricity production. Switching from coal to gas generation has largely involved a switch from base load coal generation to intermediate load[15] natural gas combined cycle (NGCC) capacity. There are varying amounts of underutilized NGCC capacity available in various regions of the United States. That underutilized capacity resulted largely from the natural gas "bubble" of the 1980s and 1990s when natural gas prices were generally about $2 per MMBTU.

Fuel cost to generate electricity is the key in the decision to switch from coal to natural gas generation. If a power plant was 100% efficient, it would take 3.412 MMBTU to generate 1 MWh of electricity. Assuming that an intermediate capacity NGCC plant is 44.78% efficient, it would consume 7.619 MMBTUs to generate 1 MWh of electricity. Similarly, a base load coal plant with an efficiency of 33.64% would use 10.412 MMBTUs to generate 1 MWh of electricity.[16] The cost of fuel to generate 1 MWh of electricity from coal has historically been lower than the cost to generate power from natural gas. The economics are now changing with coal prices rising and the drop in natural gas prices in 2011 down to levels where it now becomes more expensive to generate electricity from coal, especially when power plant efficiency is taken into account.

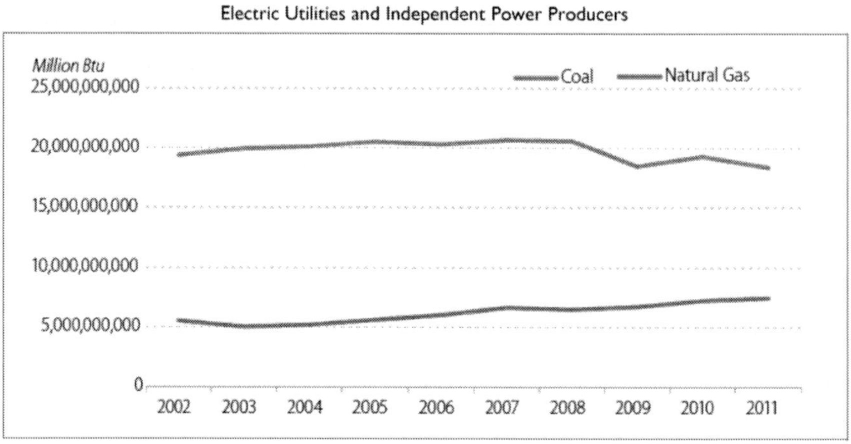

Source: CRS, data from U.S. Energy Information Administration (EIA), *Electric Power Monthly*, Table 2.1.A, "Coal: Consumption for Electricity Generation by Sector," at http://www.eia.gov/electricity/monthly/pdf/epm.pdf, September 2012; CRS, data from U.S. Energy Information Administration (EIA), Electric Power Monthly, Table 2.4.A, "Natural Gas: Consumption for Electricity Generation by Sector," at http://www.eia.gov/electricity/ monthly/pdf/epm.pdf.

Figure 3. Fuel Consumption for Electricity Generation 2002-2011.

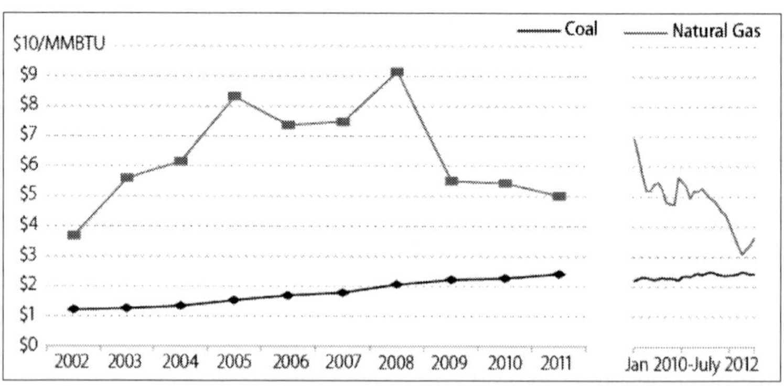

Source: CRS, data from EIA, "Table 4.2, Receipts, Average Cost, and Quality of Fuel, Electric Utilities" at http://www.eia.gov/electricity/monthly/pdf/epm.pdf. CRS. Data from EIA, Table 4.2, Receipts, Average Cost, and Quality of Fuel, Electric Utilities at http://www.eia.gov/electricity/monthly/pdf/epm.pdf.

Figure 4. Cost of Fuel for Electricity Generation, 2002-2011.

While the chart in **Figure 5** shows the expected rise and fall of power production with the seasons in the 2009 to 2012 timeframe, a look behind the numbers reveals the shift in generation as more power is being dispatched from higher efficiency natural gas-fired power plants.[17] During this time, coal's share of total U.S electricity generation has dropped and the use of natural gas to generate electricity has risen as newer, more efficient natural gas-fired generating capacity are being dispatched.[18] In April 2012, for the first time in history, the amount of electricity generation from natural gas equaled that of coal, according to EIA statistics, each with about 32% of the market.[19]

Many in the energy industry believe that a structural change in the economics of natural gas use has begun.

> While coal is projected to retain the largest share of the electricity generation mix through 2035, analyses included in the Annual Energy Outlook 2012 ... anticipate its share declining as more generation comes from natural gas and renewable technologies. Coal's role as the preeminent source of electricity generation in the United States has lessened in recent years, declining from 49% of total electricity generation in 2007 to 42% in 2011.... Projected fuel prices and economic growth are key factors influencing the future electricity generation mix. The price of natural gas, coal's chief competitor, has dropped significantly in recent years due to the increase in domestic production of natural gas.[20]

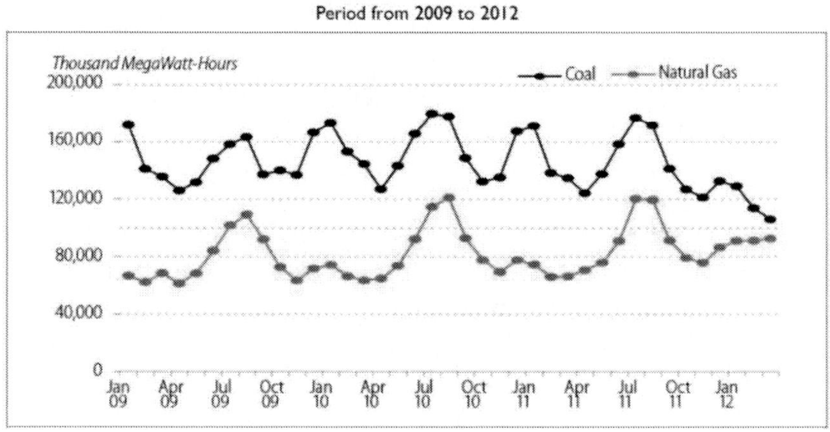

Source: CRS, data from EIA. See http://www.eia.gov/todayinenergy/detail.cfm?id=6550.

Figure 5. Net Generation—Coal vs. Natural Gas.

Nevertheless, there is still much debate as to whether the "shale gas revolution" will result in a predictable, long-term increase in supplies of natural gas at relatively low prices.

> Considerable uncertainty exists regarding the size of the economically recoverable U.S. shale gas resource base and the cost of producing those resources. Across four shale gas resource scenarios from the [EIA's] Annual Energy Outlook 2012, natural gas prices vary by about $4/MMBtu in 2035, demonstrating the significant impact that shale gas resource uncertainty has in determining future natural gas prices. This uncertainty exists primarily because shale gas wells exhibit a wide variation in their initial production rate, rate of decline, and estimated ultimate recovery per well (or EUR, which is the expected cumulative production over the life of a well).[21]

If these previously unconventional shale gas resources can be economically developed and produced in an environmentally acceptable manner, then a sustained, relatively inexpensive supply of natural gas could persist.[22]

As long as natural gas prices are "relatively" high, development of resources will continue. When a surplus of natural gas supplies causes the price to drop, resource development activities solely focused on natural gas tend to curtail. But fracking of shale gas resources is changing the outlook for natural gas supplies, and the recovery of oil from wet shale plays is currently driving development. Even so, the expectation of lower, less volatile prices from increased supplies is encouraging the greater use of natural gas for electric power generation and other industrial uses. Natural gas prices are still subject to forces of supply and demand, with the cost of developing incremental supplies determining future supplies of the resource. Further development of the interstate pipeline system may also be needed if these new natural gas supplies are to reach markets.[23]

How the natural gas industry will accommodate the expected demands of the power generation industry for gas supplies is also uncertain. Power plants purchase coal under long-term contracts, amassing the fuel on-site in large, managed coal storage piles which may hold a 30-day supply. Most natural gas power generation plants are served directly by pipelines, and manage the fuel as more of a "just-in-time" resource, since power plants typically do not have large natural gas storage capabilities. The Federal Energy Regulatory Commission observes that based on recent supply issues, this new volume of expected demand for natural gas may require more longer term coordination and planning between the electricity and natural gas supply industries.[24]

Potential Coal Plant Retirements

All power plants are subject to retirement when they reach they reach the end of their useful service life. While older coal-fired power plants are usually well-maintained, they are generally not as efficient as newer power plants. As power plants age, they are generally upgraded to continue operations, but the least efficient plants may be retired. Other plants may be shifted from base load operations (in which they essentially operate around the clock) to less demanding intermediate or peaking schedules. The cost of building a power plant is generally recovered over the depreciable life of the asset, such that operations and maintenance (O&M) expenses become the major component of an older power plant's continuing costs. A major component of O&M is the cost of fuel, and the expectation of continued lower prices for natural gas is weighing on decisions concerning whether many older, less efficient coal power plants will be mothballed or closed altogether. However, the costs of modernizing older power plants to meet new regulatory requirements can be relatively high. When the cost of upgrades to meet new environmental requirements is considered along with (perhaps increasing) O&M expenses, many older coal power plants are likely to face outright retirement decisions.

Over the last 40 years, Congress has directed EPA to reduce the potential health and environmental impacts of fossil fuel use by limiting emissions by-products or other consequences of combustion processes.[25] These environmental regulatory requirements have been evolving in the last decade due to various challenges to the regulatory implementation of federal laws. Much attention has focused recently on the resulting finalization of some of these regulations, and their potential to contribute to the retirement of some coal-burning power plants.

In addition to being the largest source of electric power, coal-fired power plants are among the largest sources of air pollution in the United States. Under the Clean Air Act (CAA),[26] however, they have not necessarily been subject to stringent requirements: emissions and the required control equipment can vary depending on the location of the plant, when it was constructed, whether it has undergone major modifications, the specific type of fuel it burns, and, to some extent, the vagaries of EPA enforcement policies. More than half a dozen separate CAA programs could potentially be used to control emissions, which makes compliance strategy potentially complicated for utilities and difficult for regulators. Because the cost of the most stringent available controls, for the entire industry, could range into the tens of billions of dollars, power companies have fought hard and rather successfully to limit or delay regulations affecting them, particularly with respect to plants constructed before the Clean Air Act of 1970 was passed. However, EPA contends that such costs would be more than outweighed by societal benefits from reduced respiratory and other illnesses—"externalities" that are largely not currently accounted for in the price of electricity generation but which are estimated by EPA as costing the United States billions of dollars annually in health care and environmental costs.[27]

Some of the new rules under development at the EPA would be implemented at the federal level, while others would be implemented at the state level. They include the Cross-State Air Pollution Rule[28] (which replaced the Clean Air Interstate Rule); the Utility Maximum Achievable Control Technology (Utility MACT) rule to reduce emissions of mercury, other metallic toxics, acid gases, and organic air toxics; the wet ash classification rule for coal combustion residues; and the Clean Water Act Section 316(b) guidelines for once-through cooling water systems. However, only the Utility MACT rule is currently in effect. EPA also proposed standards for greenhouse gas (GHG) emissions which would require all new power plants to restrict carbon dioxide emissions.[29]

EPA has yet to propose rules for GHG emissions from existing power plants, as is required by court order. Due to a general perception in the electric power industry that these new and pending environmental regulations present conflating requirements with unrealistic timeframes for compliance, the regulations have come to be referred to by the industry as the "train wreck" scenario[30] due to a perception that a negative impact on reliability could result. However, environmental groups, and some in the electric power industry—mainly those with significant investments in nuclear or natural gas-fired generation—consider the concerns overstated.[31]

In 2011, the North American Reliability Corporation (which has oversight of the bulk electric grid for reliability purposes) analyzed the effects of the proposed EPA regulations for their potential to accelerate retirement of fossil-fired units.[32] The 2011 assessment was a revision of a previous analysis to take into account "updated assumptions determined from proposed rules, as well as other viewpoints on how the rule will ultimately be carried out (i.e., how states may implement regulations), to determine potential resource adequacy impacts." NERC noted that the expected retirements of electric generators are driven by many economic factors, "not simply the cost of pollution control equipment." However, NERC's assessment estimated that between 35 GW and 59 GW of generating capacity could be potentially retired as a result of the proposed EPA rules. NERC summarized its concerns as to the possible effects on reliability:

> Existing and proposed environmental regulations in the U.S. may significantly affect bulk power system reliability depending on the scope and timing of the rule implementation and the mechanisms in place to preserve reliability.[33]

For its part, the Department of Energy (DOE) estimated in 2011 that new air pollution, water, and waste disposal proposed rules could eventually result in the retirement of between 35 GW and 70 GW of coal-fired capacity.[34]

ELECTRICITY RELIABILITY—STATE AND MARKET INPUTS

As retirement decisions are announced for growing amounts of coal capacity, questions are being asked about the cost and types of generation which may replace that capacity. EIA defines electric system reliability as "the degree to which the performance of the elements of the electrical system results in power being delivered to consumers within accepted standards and in the amount desired." Implicit in the definition is the necessity of generation adequacy to ensure that there is sufficient power plant capacity to meet projected demand, and additional reserves for exceptionally high demand periods. This includes voltage support (i.e., power injected into the grid to maintain voltage at acceptable limits) and ancillary services (i.e., services which support the reliable operation of the transmission system as it moves electricity from generating sources to retail customers) provided by retiring plants. Timely and appropriate replacement of retiring coal capacity would be essential to maintain reliability levels.

Cost recovery for a new power plant entails not only financial and performance risks, but also potential regulatory risk. In traditional cost of service states, adding a new power plant to a utility's rate base can raise customer rates.

While alternatives to potentially expensive new power plant construction may exist (such as demand response), if it appears that adding a new plant or plants to rate base will result in a significant rate increase, a reluctance may ensue on the part of state officials to authorize the addition especially in times of economic hardship. Similarly, investorowned utilities are wary of taking on too much debt for new power plant construction as this may decrease earnings and share price, possibly resulting in a downgrade of a company's financial rating.

In states with traditional cost of service ratemaking (in which electric utility rates are approved by state utility commissions), formal authorization is required before electric utilities can site, build, or place power plants into service thereby receiving recovery of costs in electricity rates. The need for new power plants is typically determined years in advance by a formal process estimating future demand and the ability of existing generation resources to meet that demand. A commonly used process is integrated resource planning (IRP), which considers a variety of supply and demand-side options, and assesses them against a common set of planning objectives and criteria, including environmental impact. The IRP objective is to help meet future customer demand by identifying the need for generating capacity, and determining the best mix of resources (including demand response) to fill the need at the lowest overall cost. Electric utilities often seek to shift the financial risk of building new power plants to ratepayers by obtaining regulatory approval to begin collecting the cost of the power plant in customer rates before the plant is in service (i.e., as construction work-in-progress).

Even in RTO markets where power generation is a competitive business, siting of new power plants is subject to state regulatory processes. However, in competitive markets, the financial risk of building a new power plant lies with the generator as they seek to bid to sell electricity into the RTO market. RTOs use markets to make operational decisions (such as generator dispatch) and to determine the provider(s) and prices for many of these services. RTOs do not own generation, but rely on price signals (based on real-time prices) to incentivize entry of new generation resources. Some RTOs employ forward capacity markets (up to three years in advance of expected needs) with auctions using either vertical demand curves (establishing quantity and allowing price to be set by the market) or sloped demand curves (regulating the price of new capacity) to contract for future capacity.

Other RTOs use an administrative approach to determine capacity requirements and set prices. However, there is some debate as to how these market approaches work, and the resulting impact on prices. Other issues include whether forward capacity markets result in new capacity actually being built, whether the appropriate type of capacity (i.e., base load vs. intermediate load) is being incentivized and attracted, or even if the administratively-set or auction-determined capacity price is above the actual cost of new entry in these markets. The issue of prevailing high electricity prices in some areas (i.e., in load pockets) has led some states in RTO regions to consider authorizing the construction of power plants or bilateral contracting for electric power independent of RTO capacity market processes.

The expected retirement of approximately 27 GW of coal-fired capacity by 2016 has been reported to EIA by coal plant owners and operators, accounting for approximately 8.5% of U.S. coal-fired capacity. While the costs of compliance with new EPA environmental regulations are a factor, several other issues are cited by coal plant owners and operators as contributing to these retirement decisions including the age of coal-fired power plants, modest electricity demand growth, the availability of previously underutilized natural gas combined-cycle power plants,[35] and the lower price of natural gas due to shale gas development. [36] And some coal plants which have made significant modifications to meet existing EPA regulations are being closed or mothballed due to a combination of low natural gas prices, and either transmission congestion[37] or a lack of transmission options to sell power into other markets.[38]

The implications of coal-fired electric power plant retirements will likely play out on a regional basis, as EPA's rules go through inevitable challenges and modifications as the regulations are finalized.[39] Power generation in the United States varies according to local energy resources, and cost and availability of infrastructure for fuels transportation.[40]

The Eastern (except New England) and Midwest regions of the United States are more coal-dependent than the Northeast or Western regions, as shown in **Figure 6**.

Several large regional transmission organizations[41] (RTOs) have undertaken studies of the impact of the new EPA rules. A major question being asked in these studies is whether there will be a risk to reliability based on the timing of requirements, and the scheduling and availability of equipment and personnel to accomplish retrofits of power plants to bring them into compliance with EPA regulations. The results of a study[42] by the Midwest ISO (MISO), the RTO serving 11 states and one Canadian province may provide insights into the implications of both EPA regulations and lower natural gas prices.

There is approximately 70 GW of coal-fired capacity in the MISO market,[43] with an average coal plant age of 45 years.[44] The study identified potential retirements because of proposed EPA rules, a low natural gas price, and various levels of carbon mitigation costs. MISO summarized the results of its study noting that EPA regulations and lower natural gas prices would likely result in between 3 GW and 13 GW of coal capacity retirements, with MISO itself likely to see costs of between $32 billion and $33 billion due to EPA compliance retrofit costs, new capacity charges, system energy costs, and transmission system upgrades to maintain reliability.[45]

MISO does not own or build power plants. State utility commissions still control the process of new power plant construction in many states. States often require a utility to use integrated resource planning methods[46] to address system needs. If a determination is made that new generating capacity is needed to ensure reliability, then how to requisition that capacity becomes the responsibility of state public utility commissions or the RTO markets. The decision would likely be made according to how they view the effects of coal plant retirements on the levels and types (i.e., base load vs. intermediate or peak load power plants) of capacity that will be needed to serve electricity demands.[47] RTO markets generally rely on price signals[48] to provide an incentive for new generation to enter the market, or employ forward capacity markets with higher prices to encourage new generation to be built. Other "generation-like" resources (such as demand response,[49] energy efficiency, and conservation) would also likely be employed to reduce the need for building new capacity.

Given that significant underutilized NGCC exists in various U.S. regions, the possibility of further shifting from coal base load plants to natural gas intermediate capacity exists. A recent study[50] by the Massachusetts Institute of Technology in 2011 noted that the existing U.S. NGCC generation fleet had an average capacity factor[51] of approximately 41%, while its design capacity allowed such plants to operate at 85%. The MIT study looked at a scenario across selected regions of the United States which mimicked the "full dispatch" of existing natural gas combined cycle plants. The study concluded that under such a scenario (while noting that transmission constraints exist), there is "sufficient surplus NGCC capacity to displace roughly one-third of U.S. coal generation, reducing CO2 emissions from the power sector by 20%."[52]

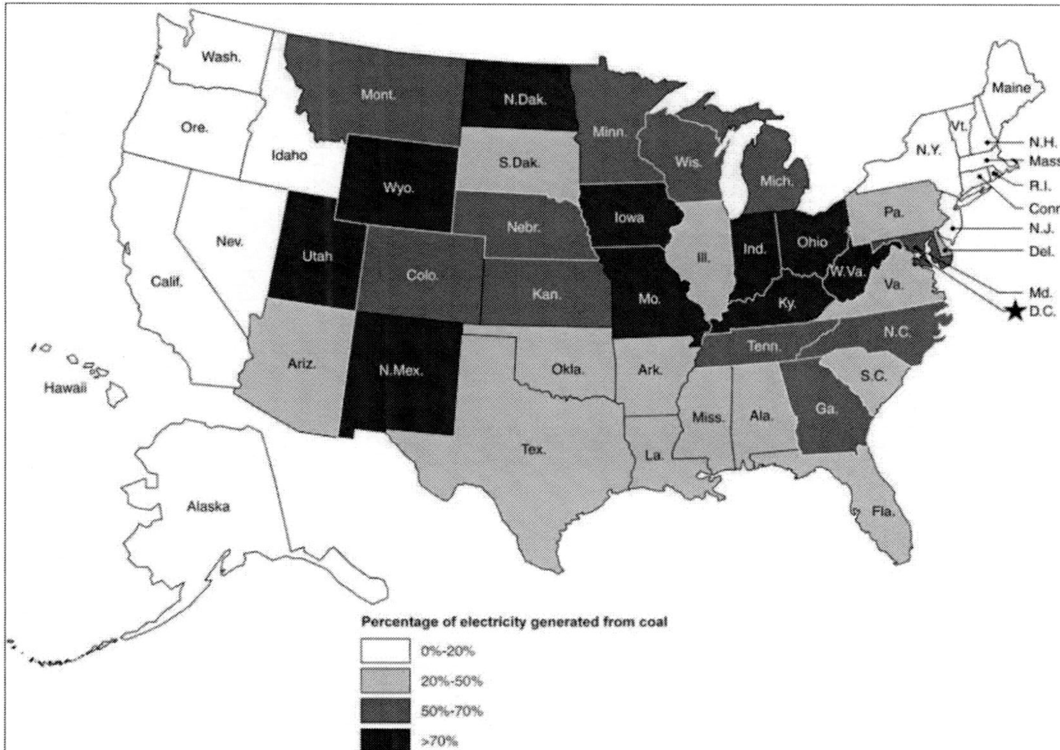

Source: Government Accountability Office. See http://www.gao.gov/assets/600/ 592542.pdf.

Figure 6. Percentage of Electricity Generated from Coal by State, 2010.

Role of New and Alternative Power Generation Technologies

According to the EIA, coal is expected to be a key part of electricity generation in the United States well past the year 2030.[53] Electric utilities value diversity in power generation options since reliance on one fuel or technology can leave them vulnerable to price and supply volatility. However, the choices for new power plant technologies will likely be subject to proposed EPA rules limiting GHG emissions. The proposed rule published under the Clean Air Act's New Source Performance Standards program would limit emissions from new power plants to no more than 1,000 pounds of carbon dioxide for each megawatt-hour (MWh) of electricity produced. Today's NGCC generation emits on average about 800 pounds of carbon per MWh compared to current U.S. coal plants with average emissions of about 1,800 pounds of carbon dioxide per each MWh of electric power produced.[54] The proposed GHG rule would allow an exception for the approximately one dozen new coal plants that are permitted and beginning construction within one year, and would not apply to existing plants. Only new coal plants with carbon capture and storage (CCS) are likely to meet the proposed GHG regulations,[55] and CCS technology is estimated to add about 30% to a coal plant's cost. However, according to EIA, there are no new conventional coal plants expected to come on line after 2012; 2 GW of coal plant capacity with CCS is expected to enter service by about 2017.[56] New advanced NGCC plants entering the market in 2017 are expected to cost approximately $63 per MWh, while new advanced coal plants are estimated to cost $111 per MWh, and advanced coal with CCS is estimated at $139 per MWh.[57] Based on levelized cost estimates alone, new coal generation (even without CCS) may not be competitive with new NGCC given the built-in expectation of future fuel costs in the estimates.

The variability of most renewable electricity technologies means that these technologies offer limited opportunities for replacing coal generation, and these opportunities vary across the United States. Given that the best wind and solar resources are located far from population centers, opportunities to use wind power and solar power are likely dependent on development of transmission lines. Without major breakthroughs in technology efficiency or energy storage, growth of power generation from all renewable sources[59] is expected to be limited to no more than 16% of all generation by 2035.[60]

> Since load must be balanced on a continuous basis, units whose output can be varied to follow demand (dispatchable technologies) generally have

more value to a system than less flexible units (non-dispatchable technologies) or those whose operation is tied to the availability of an intermittent resource.[61]

Development of joint renewable electricity and natural gas-fired power projects as "hybrid generation" is a possibility in certain regions. This could make variable renewable resources like wind and solar power more dispatchable, and would likely be dependent on state renewable portfolio standard[62] requirements.

CARBON CAPTURE AND SEQUESTRATION (CCS)[58]

Carbon capture and sequestration (or storage)—known as CCS—has attracted congressional interest as a measure for mitigating global climate change. Large amounts of carbon dioxide (CO_2) emitted from fossil fuel use in the United States are potentially available to be captured and stored underground using CCS, and prevented from reaching the atmosphere. Large, industrial sources of CO_2, such as coal-fired electricity-generating plants, are likely initial candidates for CCS because they are predominantly stationary, single-point sources. Electricity generation contributes over 40% of U.S. CO_2 emissions from fossil fuels. Currently, U.S. power plants do not capture large volumes of CO_2 for CCS.

To date, there are no commercial ventures in the United States that capture, transport, and inject industrial-scale quantities of CO_2 solely for the purposes of carbon sequestration. In 2012, the U.S. Department of Energy's CCS research, development, and deployment program embarked on commercial-scale demonstration projects for CO_2 capture, injection, and storage. The success or failure of these projects will likely bear heavily on the future outlook for widespread deployment of CCS technologies as a strategy for preventing large quantities of CO_2 from reaching the atmosphere while U.S. power plants continue to burn fossil fuels, mainly coal.

Nuclear power may be able to take over more of the base load generation role from coal, but the path forward for new nuclear generating units in such a role is uncertain. According to the Nuclear Energy Institute, as of 2012, five new nuclear power units are under construction, and maintaining nuclear energy's current share of 20% of total generation would require "building one reactor every year starting in 2016, or 20 to 35 new units by 2035."[63] The cost of building new nuclear power units may present a barrier to this goal,[64] as might concerns raised by the nuclear incident at Fukushima.

Given current technological options for power generation in the United States, the long-term expectation of lower natural gas prices in combination with existing and proposed EPA rules suggests that the option of coal use as a fuel for electric power generation faces an uncertain future.

INDUSTRIAL COAL CONSUMPTION[65]

In 2011 approximately 72 Million Short Tons (MST) of coal were used by the U.S industrial sector, which represented roughly 7% of total coal consumption in the country.[66] The industrial sector's share of U.S. coal demand has declined substantially since 1949, when the industrial sector accounted for nearly 45% of annual coal consumption (see **Figure 7**).[67] The decline of the industrial sector's share of coal consumption is generally a result of three developments since 1949: (1) reduced coal use for coke production and steel making, (2) reduced coal use for industrial process applications, and (3) increased coal use for electric power production, which now represents more than 90% of U.S. coal demand.[68]

Coal has a number of industrial uses including steel making, cement production, combined heat and power (CHP), and process heating. Additionally, coal is used by a number of industrial operations including alumina refineries, chemical producers, pharmaceutical companies, and paper manufacturers.[69] According to the World Coal Institute, there are thousands of products that include coal or coal by-products: soap, aspirins, solvents, dyes, plastics, fibers, and specialty products (e.g., activated carbon, carbon fiber, and silicon metal).[70]

The Energy Information Administration classifies industrial sector coal consumption into three separate categories: (1) Coke Plants, (2) Combined Heat and Power (CHP), and (3) Non-CHP. **Table 1** indicates the amount of coal consumed in 2011 for each of these categories.

Table 1. 2011 Industrial Sector Coal Consumption (Million Short Tons)

Coke Plants	CHP	Non-CHP	Total
21.4	24.7	25.6	**71.7**

Source: Energy Information Administration.

(1949-2011)

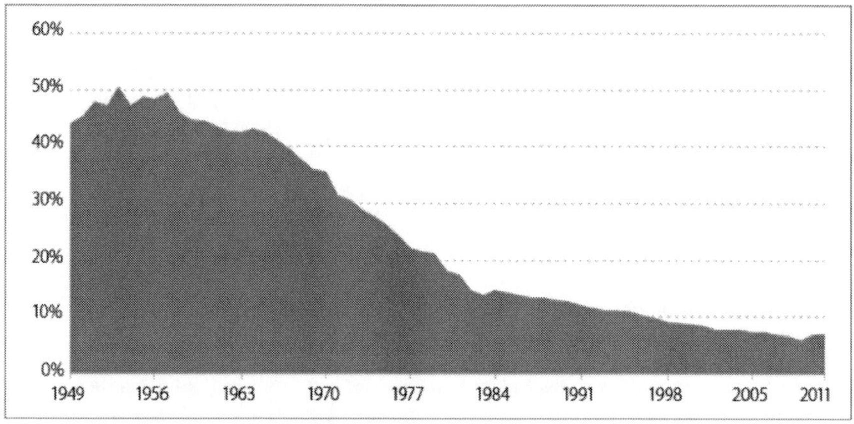

Source: Energy Information Administration.

Figure 7. Industrial Sector Share of Total U.S. Coal Consumption.

ISSUES FOR CONGRESS

Given the expected increase in natural gas supplies and expected shift in electricity production from coal to natural gas as a primary fuel, the issue of what level of coal generation should be preserved for power generation in the United States may be a question Congress will be faced with. While U.S. coal currently used for power generation may have options for markets overseas, the domestic future of coal power generation is clouded by both market prices and environmental regulations. The United States is believed to have a bountiful supply of coal. But how much or whether the future of coal will include power generation may be a question of national energy policy.

Congress may be also confronted with the question of whether an "all-of-the above" strategy for energy is viable. The electric utility industry values diversity in fuel choice options since reliance on one fuel or technology can leave electricity producers vulnerable to price and supply volatility. However, an "inverse relationship" is developing for coal vs. natural gas as a power generation choice based on market economics alone, and policies which allow one fuel source to dominate may come to the detriment of the other.

Congress may also face a decision as to whether coal-fired power generation can be a viable technology option for the future. New advanced NGCC plants entering the market in 2017 are expected to cost approximately $63 per MWh, while new advanced coal plants are estimated to cost $111 per

MWh, and advanced coal with CCS is estimated at $139 per MWh. The cost difference between these technologies, and proposed GHG emissions limits, raise concerns over coal's future as a viable fuel choice for power generation. It is not clear (at this point in time) whether research and development will result in new technologies able to reduce the differential.

End Notes

[1] The heating value of any fuel is the energy released per unit mass or per unit volume of the fuel when the fuel is completely burned. Higher heat value fuels liberate more energy per unit of mass or volume.

[2] A simple cycle natural gas power plant burns natural gas in a combustion turbine (like a jet engine) to turn a generator which produces electricity. When used in a combined cycle mode, the hot exhaust gases from the combustion turbine are used to generate steam to turn a steam turbine to efficiently create additional electricity.

[3] One billion watts or one thousand megawatts.

[4] See http://www.eia.gov/todayinenergy/detail.cfm?id=7290.

[5] See http://www.eia.gov/totalenergy/data/annual/showtext.cfm?t=ptb0811a.

[6] A corporation, person, agency, authority, or other legal entity or instrumentality that owns or operates facilities for the generation of electricity for use primarily by the public, and that is not an electric utility. See http://www.eia.gov/tools/glossary/index.cfm?id=I#ind_pwr_prod.

[7] EIA, *Table 2.1.A. Coal: Consumption for Electricity Generation*, 2002-2011, http:// www.eia.gov/electricity/monthly/pdf/epm.pdf.

[8] A plant, usually housing high-efficiency steam-electric units, which is normally operated to take all or part of the minimum load of a system, and which consequently produces electricity at an essentially constant rate and runs continuously. These units are operated to maximize system mechanical and thermal efficiency and minimize system operating costs. EIA, *EIA Glossary "Base Load Plant"*, http://www.eia.gov/tools/glossary/index.cfm.

[9] A plant usually housing old, low-efficiency steam units, gas turbines, diesels, or pumped-storage hydroelectric equipment normally used during the peak-load periods. See http://www.eia.gov/tools/glossary/index.cfm.

[10] Black and Veatch, *Cost Report—Cost and Performance Data for Power Generation Technologies*, National Renewable Energy Laboratory, February 2012, http://bv.com/docs/reports-studies/nrel-cost-report.pdf.

[11] Typically, all the freight cars on a "unit train" are filled with coal being shipped directly to a single destination.

[12] See http://www.eia.gov/coal/transportationrates/.

[13] See ISO New England, "Scenario Analysis Project—Long Term Forecast of Oil, Natural Gas and Coal Prices in New England," 2007, http://www.isone.com/committees/comm_wkgrps/othr/sas/mtrls/apr22007/fuel-price-forecast.pdf.

[14] See Energy Information Administration, Office of Oil and Gas, *An Analysis of Price Volatility in Natural Gas Markets*, "Figure 1: Daily Henry Hub Spot Price, 1994-2006," August 2007, http://www.eia.gov/pub/oil_gas/natural_gas/feature_articles/2007/ngprivolatility/ngprivolatility.pdf.

[15] The range from base load to a point between base load and peak load.

[16] Sandy Fielden, "The Economics of Coal-to-Gas Switching," *Power Magazine*, August 21, 2012, http://www.powermag.com/business/The-Economics-of-Coal-to-Gas-Switching_4880_p4.html.

[17] EIA, "Natural Gas Spot Prices near 10-Year Lows Amid Warm Weather and Robust Supplies," February 1, 2012, http://www.eia.gov/todayinenergy/detail.cfm?id=4810.

[18] "Coal generation decreased 29 billion kilowatt-hours from March 2011 to March 2012, while natural gas generation increased 27 billion kilowatt-hours during the same time period. In March 2012, coal's share of total generation was34% compared to natural gas at 30%." See EIA, "U.S. Coal's Share of Total Net Generation Continues to Decline," June 5, 2012, http://www.eia.gov/todayinenergy/detail.cfm?id=6550.

[19] U.S. Energy Information Administration, *Net Generation by Energy Source: Total (All Sectors), 2002-June 2012*, July 2012, http://www.eia.gov/electricity/monthly/epm_table_grapher.cfm?t=epmt_1_1.

[20] EIA, "Fuel Used in Electricity Generation Is Projected to Shift over the Next 25 Years," July 30, 2012, http://www.eia.gov/todayinenergy/detail.cfm?id=7310.

[21] EIA, Projected Natural Gas Prices Depend on Shale Gas Resource Economics," August 27, 2012, http://www.eia.gov/todayinenergy/detail.cfm?id=7710.

[22] For further discussion of natural gas production, see CRS Report R42814, *Natural Gas in the U.S. Economy: Opportunities for Growth*, by Robert Pirog and Michael Ratner

[23] Jack Buehrer, *Natural Gas Boom Drives Pipeline Upgrades*, Engineering News Record, May 7, 2012, http://enr.construction.com/infrastructure/power_industrial/2012/0507-shale-boomhttpprimisphmsadotgovconstructionconstphmsahtm.asp.

[24] Federal Energy Regulatory Commission, *Natural Gas - Electric Coordination*, August 2012, http://www.ferc.gov/industries/electric/indus-act/electric-coord.asp.

[25] For example, with the Clean Air Act Amendments of 1970 (P.L. 91-604) and subsequent revisions, the Clean Water Act of 1972 (P.L.92-500), and the Resource Conservation and Recovery Act of 1976 (P.L. 94-580).

[26] The Clean Air Act, codified as 42 U.S.C. 7401 et seq., seeks to protect human health and the environment from emissions that pollute ambient, or outdoor, air. It requires the Environmental Protection Agency to establish minimum national standards for air quality, and assigns primary responsibility to the states to assure compliance with the standards. Areas not meeting the standards, referred to as "nonattainment areas," are required to implement specified air pollution control measures. The Act establishes federal standards for stationary and mobile sources of air pollution and their fuels and for sources of 187 hazardous air pollutants, and it establishes a cap-and-trade program for the emissions that cause acid rain. It establishes a comprehensive permit system for all major sources of air pollution. It also addresses the prevention of pollution in areas with clean air and protection of the stratospheric ozone layer. For more information, see CRS Report RL30853, *Clean Air Act: A Summary of the Act and Its Major Requirements*, by James E. McCarthy, Claudia Copeland, and Linda-Jo Schierow.

[27] USEPA Office of Air and Radiation, *The Benefits and Costs of the Clean Air Act: 1990 to 2020*, August 2010, http://www.epa.gov/air/sect812/aug10/fullreport.pdf.

[28] On August 21, 2012, in a 2-1 decision, the court vacated and remanded the rule, finding that EPA's imposition of Federal Implementation Plans, without first giving the states an opportunity to develop their own plans, was unlawful. See CRS Report R41563, *Clean Air Issues in the 112th Congress*, by James E. McCarthy. By vacating the rule, the Court determined that the Clean Air Interstate Rule would still apply. But that rule was itself vacated, making future implementation of any air transport rule unclear.

[29] Environmental Protection Agency, *Carbon Pollution Standard for New Power Plants*, http://epa.gov/carbonpollutionstandard/.

[30] These regulations are discussed in separate CRS Reports: CRS Report R41914, *EPA's Regulation of Coal-Fired Power: Is a "Train Wreck" Coming?*, by James E. McCarthy and Claudia Copeland, and CRS Report R42570, *Proposals to Amend RCRA: Analysis of Pending Legislation Applicable to the Management of Coal Combustion Residuals*, by Linda Luther.

[31] CRS Report R42144, *EPA's Utility MACT: Will the Lights Go Out?*, by James E. McCarthy.

[32] North American Electric Reliability Corporation, *Potential Impacts of Future Environmental Regulations—Extracted from the 2011 Long-Term Reliability Assessment*, November 2011, http://www.nerc.com/files/EPA%20Section.pdf.

[33] Ibid.

[34] Debra McCown, "EPA Regulations for Coal-Fired Power Plants Could Force Shut Downs," *Bristol Herald Courier*, May 24, 2011, http://www2.tricities.com/business/2011/may/25/wood-gives-dire-warning-due-epa-regulations-coal-f-ar-1062322/.

[35] An electric generating technology in which electricity is produced from otherwise lost waste heat exiting from one or more gas (combustion) turbines. The exiting heat is routed to a conventional boiler or to a heat recovery steam generator for utilization by a steam turbine in the production of electricity. This process increases the efficiency of the electric generating unit. See http://www.eia.gov/tools/glossary/index.cfm.

[36] EIA, *27 GigaWatts of Coal-Fired Capacity to Retire over Next Five Years*, July 27, 2012, http://www.eia.gov/todayinenergy/detail.cfm?id=7290#.

[37] A condition that occurs when insufficient transfer capacity is available to implement all of the preferred schedules for electricity transmission simultaneously. See http://www.eia.gov/tools/glossary/index.cfm.

[38] Gib Snyder, "An Inside Look at the Process for Mothballing," *The Observer*, September 9, 2012, http://www.observertoday.com/page/content.detail/id/575844/An-inside-look-at-the-process-for-mothballing.html?nav=5047.

[39] InsideEPA, *EPA Urges Appellate Court To Rehear Ruling Vacating Cross-State Air Rule*, October 5, 2012, http://insideepa.com/201210052412215/EPA-Daily-News/Daily-News/epa-urges-appellate-court-to-rehear-ruling-vacating-cross-state-air-rule/menu-id-95.html.

[40] "In New England, most of the coal that will be burned … will likely originate in South America or Indonesia, where the chemical composition of the coal is less harmful to the environment. Western U.S. coals also have a favorable chemical composition, but it is currently too expensive to deliver coal from the Rocky Mountains to New England relative to foreign supplies." ISO New England, *Scenario Analysis Project—Long Term Forecast of Oil, Natural Gas and Coal Prices In New England*, 2007, http://www.isone.com/committees/comm_wkgrps/othr/sas/mtrls/apr22007/fuel-price-forecast.pdf.

[41] The purpose of a Regional Transmission Organization is to ensure the efficient and reliable delivery of power across large areas. RTOs facilitate and promote efficiency in wholesale electricity markets and ensure that electricity consumers pay the lowest price possible for reliable service by removing transmission barriers between buyers and sellers. See http://www.pjm.com/about-pjm/learning-center/pjm-overview/energy-basics/what-is-a-rto.aspx.

[42] MISO, *EPA Impact Analysis- Impacts from the EPA Regulations on MISO*, October 2011, https://www.midwestiso.org/Library/Repository/Study/MISO%20EPA%20Impact%20Analysis.pdf. (MISO EPA)

[43] Of the 70,000 MW of coal-fired capacity in the MISO market, less than half does not have plans for SO2 controls. Furthermore, 38% have no SO2 controls or NOx controls, and 38% have no SO2 controls or Fabric Filters. (MISO EPA)

[44] The life expectancy of coal-fired power plants from a cost recovery basis is generally estimated to be 40 years.

[45] MISO EPA.

[46] "Integrated resource planning [IRP] is a crucial element for success in a constantly changing business and regulatory environment and is based on comprehensive, holistic and risk aware analysis. The integrated approach considers a broad spectrum of feasible supply and demand-side options and assesses them against a common set of planning objectives and criteria, including environmental impact. The IRP objective is to help meet future customer demand by identifying the need for generating capacity and determining the best mix of resources to fill the need. The capacity gap is the difference between the projected firm (or known) requirements and existing firm supply." See TVA, *TVA's Environmental and*

Energy Future, Integrated Resource Plan, March 2011, p. 27, http://www.tva.gov/environment/reports/irp/pdf/Final_IRP_Ch1.pdf.

[47] Many electric utilities are required by states to undertake long-range planning. "Integrated resource planning helps electric utilities choose the best resource options to generate electricity and other options to meet customer expectations for energy services. Increasing competition, changing technologies, and environmental concerns are among the many issues utilities must consider when developing their plans." See "Purpose of and Need for Integrated Resource Planning" at http://www.tva.gov/environment/reports/energyvision2020/ev2020_vol1ch01.pdf.

[48] Shortages of capacity cause electricity prices to rise in a competitive market.

[49] Demand response (also known as load response) is end-use customers reducing their use of electricity in response to power grid needs, economic signals from a competitive wholesale market or special retail rates. See http://www.pjm.com/markets-and-operations/demand-response.aspx.

[50] Massachusetts Institute of Technology, *The Future of Natural Gas*, 2011, http://mitei.mit.edu/system/files/NaturalGas_Report.pdf. (MIT Study)

[51] The ratio of the electrical energy produced by a generating unit for the period of time considered to the electrical energy that could have been produced at continuous full power operation during the same period. Seehttp://www.eia.gov/tools/glossary/index.cfm.

[52] MIT Study, op. cit.

[53] "In the Reference case, the natural gas share of electric power generation increases from 24 percent in 2010 to 28 percent in 2035, while the renewables share grows from 10 percent to 15 percent. In contrast, the share of generation from coal-fired power plants declines. The historical reliance on coal-fired power plants in the U.S. electric power sector has begun to wane in recent years. Over the next 25 years, the share of electricity generation from coal falls to 38 percent, well below the 48-percent share seen as recently as 2008, due to slow growth in electricity demand, increased competition from natural gas and renewable generation, and the need to comply with new environmental regulations." See http://www.eia.gov/forecasts/aeo/pdf/0383(2012).pdf.

[54] 40 C.F.R. § Part 60. See http://epa.gov/carbonpollutionstandard/pdfs/20120327proposal.pdf.

[55] "New coal-, coal refuse-, oil- and petroleum coke-fired boilers and IGCC [Integrated Gasification Combined Cycle] units should also be able to meet this standard by employing carbon capture and storage (CCS) technology." Ibid.

[56] Diane Kearney, *Coal Projections from the Annual Energy Outlook 2011 Early Release*, EIA, April 6, 2011, http://www.stb.dot.gov/stb/docs/RETAC/2011/April/EIA%20annual%20energy%20outlook%202011.pdf.

[57] Total levelized system cost. "Levelized cost is often cited as a convenient summary measure of the overall competiveness of different generating technologies. It represents the per-kilowatthour cost (in real dollars) of building and operating a generating plant over an assumed financial life and duty cycle. Key inputs to calculating levelized costs include overnight capital costs, fuel costs, fixed and variable operations and maintenance (O&M) costs, financing costs, and an assumed utilization rate for each plant type. The importance of the factors varies among the technologies." See EIA, *Levelized Cost of New Generation Resources in the Annual Energy Outlook 2012*, Annual Energy Outlook 2012, July 12, 2012, http://www.eia.gov/forecasts/aeo/electricity_generation.cfm. (EIA NewGen)

[58] See CRS Report R42532, *Carbon Capture and Sequestration (CCS): A Primer*, by Peter Folger, and CRS Report R42496, *Carbon Capture and Sequestration: Research, Development, and Demonstration at the U.S. Department of Energy*, by Peter Folger.

[59] Includes conventional hydroelectric, geothermal, wood, wood waste, all municipal waste, landfill gas, other biomass, solar, and wind power.

[60] EIA, *Annual Energy Outlook 2012, Table A9*, DOE/EIA-0383(2012), May 2012, http://www.eia.gov/forecasts/aeo/pdf/0383%282012%29.pdf.EIA, Table A9, Electric Generating Capacity. Annual Energy Outlook.

[61] EIA NewGen.

[62] Renewable Portfolio Standards are state programs requiring a percent of energy sales or installed capacity to come from renewable resources.

[63] Nuclear Energy Institute, *New Nuclear Energy Facilities*, 2012, http://www.nei.org/keyissues/newnuclearplants/.

[64] The two units being built by the Southern Company at the Vogtle station in Georgia were originally estimated to cost $14 billion. See Ray Henry, "Building Costs Rise at U.S. Nuclear Sites," *South Coast Today.com*, July 15, 2012,http://www.southcoasttoday.com/apps/pbcs.dll/article?AID=/20120715/NEWS/207150345/-1/NEWSMAP.

[65] This section was written by Phillip Brown.

[66] Energy Information Administration, *Annual Energy Review*, September 27, 2012.

[67] Ibid.

[68] Ibid.

[69] "The Coal Resource: A Comprehensive Overview of Coal," World Coal Institute, 2009.

[70] Ibid.

INDEX

A

B

C

D

E

F

G

N

O

P